I0076217

De-Extinctions. A Quick Immersion

Quick Immersions uses intellectual rigor and easy language to offer a good introduction, or deeper knowledge, on diverse issues, as well-structured texts by prestigious authors delve into the worlds of political and social sciences, philosophy, science and the humanities.

Carles Lalueza-Fox

DE-EXTINCTIONS

A Quick Immersion

Tibidabo Publishing

Copyright © 2018 by Carles Lalueza-Fox.
Published by Tibidabo Publishing, Inc.

All rights reserved. No part of this publication may be reproduced, stored in a retrieval system, or transmitted, in any form or by any means, electronic, mechanical, photocopying, recording, scanning or otherwise, without the prior permission in writing of the Published, or as expressly permitted by law, or under terms agreed with the appropriate reprographics rights organization.

Translation by Lori Gerson
Cover art by Raimon Guirado
For photograph credits, please see page 7.

Published 2019

Visit our Series on our Web:
www.quickimmersions.com

ISBN: 978-1-949845-00-6
2 3 4 5 6 7 8 9 10

Printed in the United States of America.

Original title "*Des-Extinciones. Una inmersión rápida*" by Carlos Lalueza-Fox.
Este libro ha recibido una ayuda a la edición del Ministerio de Cultura y Deporte.

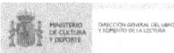

Contents

List of illustrations, tables and graphics

Introduction

In one of the most famous passages from the Bible, God decides to put an end to his own creation, disgusted by the wickedness shown by man. But he asks Noah to build an enormous ark – for which he gives him curiously precise instructions – so he can save himself and his family and some specimens from each living being. As Genesis is really an imperfect fusion of two stories, we find certain discrepancies in some passages; God orders Noah to bring two – male and female – from each species, but in the following paragraph it has become seven pairs (without a doubt a better strategy to conserve some genetic diversity). Later, he hurls down on earth a universal flood that exterminates "every living creature from the face of the earth, from mankind to livestock, snakes and to the birds in the sky."

In November 1872, when George Smith (1840-1876) was examining various tablets with cuneiform writing that had come from Nineveh in ancient Assyria, he came across one in which he was able to decipher a similar legend, which is known as The Epic of Gilgamesh. Smith began ripping his clothing as he jumped for joy around the room (alas this great discovery did not bring him good luck, given that he died within four years after returning in ill health from a disastrous excavation in the ruins of Nineveh). The Epic of Gilgamesh, which outraged Victorian society, takes place around 4,000 years ago, long before the first preserved texts of the Bible, and it is believed the Bible may have taken inspiration from it. In fact, similar legends exist among many peoples, including Norse, Mayan and Greek mythology, and even mythology from various tribes in New Guinea and North America. Undoubtedly, all these episodes are inspired on the memory of catastrophic local events. But regardless of their true scope, the underlying idea is extremely powerful: erase everything to begin anew (in the same way that, at times, with the use of x-rays hidden pictures are discovered underneath current canvases by famous painters such as Goya, Van Gogh or Picasso, who simply decided it was better to cover the work and paint on top of it).

With the discovery of bones from extinct, at times enormous, animals in areas of Europe during the Enlightenment, catastrophism became an idea with scientific entity. In France, Georges Cuvier (1769-1832) founded the area of paleontology and comparative anatomy and expanded Linnaean taxonomy to animals from the past. He described species such as the mastodon, the megatherium and the pterodactyl starting from a study of their fossil remains. Although Cuvier is saddled with a bad reputation for having attributed the disappearance of these animals to repeated catastrophic episodes and his book *Recherches sur les ossemens fossiles* had the aim of scientifically demonstrating the authority of the Bible, this biased perception seems to stem from an inaccurate introduction by Robert Jameson, a professor of natural history in Edinburgh during the era and translator of the work to English. It is true that Cuvier did not believe in evolution, but his approach to studying animals from the past was strictly scientific, and he did not discard that there had been changes and diversification of species after each catastrophe. In one of his conferences on the practice of vertebrate paleontology at the *Institut National*, he tells us that the idea is to "in some way reconstruct the skeletons of the animals, and then compare these beings that have been brought back to life with those that naturalists have discovered alive on the surface of our current planet to determine their similarities and their differences." Without a doubt, a correct definition of the discipline.

Catastrophism was discredited by the father of modern geology, Charles Lyell (1797-1875), who contributed to the emergence of stratigraphic observation and, with this, the idea that changes are small and constant over vast periods of time (which came to be known as uniformitarianism). Lyell was a friend of the young Charles Darwin, and the ideas of the latter influenced his concept of evolution as being the accumulation of gradual changes (although Lyell was never in agreement with the role of natural selection as an engine in these processes).

Tree of Life

Darwin used the metaphor of the "tree of life" – which had associated biblical connotations from Genesis – to describe the relationships that join all organisms on the planet, those that exist and those that have existed. In fact, a diagram that shows the diversification of a several species over time in successive diverse branches is the only illustration in *On the Origin of Species* (nevertheless, experts in phylogeny prefer the evolutionary tree Darwin drew with his own hand in 1837 and which appears in his notes, as it allows for the existence of some branches that are shorter than others). In a famous passage from his book, Darwin tells us: "The affinities of all the beings of the same class have sometimes been represented by a great tree. I believe this simile largely speaks the truth. The green and budding twigs may represent existing species; and those produced during each former year may represent the long succession of extinct species." This powerful image in the shape of a tree, used to imagine the history of life, immediately became an icon as it illustrated very intuitively evolutionary success and failure. In regard to the latter, Darwin emphasized in another paragraph: "… I believe it has been with the great Tree of Life, which fills with its dead and broken branches the crust of the earth, and covers the surface with its ever branching and beautiful ramifications."

Ernst Haeckel (1834-1919), a German follower of Darwin – who, by the way, coined such terms as "ecology," "phylogeny" and even "First World War" – took charge of popularizing the image of the tree through illustrations that he himself made (the most famous of which dates to 1874), with every group of organisms on a different branch and human beings on the highest branch of the treetop. However, representation in the form of a tree presents diverse conceptual problems, one of which – the existence of sap transported up and down the trunk, on a road that evolution follows only upwards – had already been detected by Darwin himself, who considered that perhaps a more fitting image would have been living corals, who would be growing on top of previous dead corals. Another of the limitations of the tree of life can be sensed in Haeckel's representations, and it is the fact that, by showing only the living branches, the erroneous impression is created that the past was less diverse and complex than the present. But it is absolutely not like that.

In fact, the amount of dead branches on our tree is not small, which offers a rather dark contrast to the bucolic description of the botanical metaphor from Darwin. It is calculated that around 99.67% of the species that have at some time lived on the earth are now extinct. As of today, biologists have discovered nearly 1.9 million live species on the earth, and it is estimated that there are various millions more to be discovered, up to a total of perhaps 8.7 (without counting the complex bacterial universe, which is a world apart). But it is estimated that extinct species surpass current species in number by 100-to-one and even 1,000-to-one. Following the model of the tree, this would be nearly impossible to represent, given the dense foliage from all these basal branches.

It could be argued that the process of extinction is as natural as that of speciation. That is, the emergence of new species who bring novel adaptive responses to new environmental challenges. This is because the earth and life are not static entities, but rather dynamic and changing. Nevertheless, it is true that extinctions are not distributed uniformly, but rather throughout earth's geological history have grouped together in five large episodes, known as mass extinctions (some argue, as we will see, that we find ourselves in the sixth of these events). A mass extinction can be defined as the disappearance of many species in a short period of time (from one to five million years or even less). To be more precise, what happens is that there is a rise in the "normal" extinction rate or a decrease in the speciation rate, or both phenomena at the same time. This can be defined in various ways, but normally it is shown as the percentage of families or species that become extinct in respect to the total present up to that moment.

Five Mass Extinctions

Since the nineteenth century, paleontologists have confirmed that there have been five mass extinctions: End – Ordovician, End – Devonian, End – Permian, End – Triassic and End – Cretaceous. The causes are variable but in general have to do with global climate, with the probable exception of the last event, which saw extinction of the dinosaurs due to the impact of a huge asteroid in the Yucatan that formed the gigantic Chicxulub crater, recently surveyed at more than one thousand meters below the surface. Other causes proposed at different events include global warming due to mass episodes of volcanism, with possible evidence of more impacts from asteroids, at least in the Permian. In general, these phenomena trigger a rise or fall in the concentration level of carbon dioxide in the air or water, which is ultimately the key factor that sparks the ecological crisis. With the possible exception of the Cretaceous episode, mass extinctions seem to respond to relatively long-term processes; even in the case mentioned, there is evidence that many evolutionary lineages had been declining for a long time. Perhaps the longest was the Devonian, which according to some researchers lasted a period of some 20 – 25 million years and had up to a dozen successive occurrences of extinction.

The Ordovician extinction ended around 443 million years ago, and in a period that lasted between 1.9 – 3.3 million years nearly 86% of species became extinct, many of them paradigmatic organisms such as trilobites, one third of all brachiopods, conodonts and many graptolites. The Devonian extinction ended around 359 million years ago and nearly 75% of the species at that time were lost, among them again many brachiopods, ammonites, trilobites, graptolites and crinoids; the coral reefs formed of stromatoporoidea sponges were especially affected, and with them a large part of the benthic fauna that lived in them. These ecosystems did not recover until the appearance of new types of corals one hundred million years later. The Permian collapse was, without a doubt, the most catastrophic and is known for good reason as "the Great Dying;" it ended 252 million years ago and in scarcely 100,000 years no less than 96% of species were lost, among them the majority of ammonites, a large part of the big amphibian labyrinthodonts and the large synapsid reptiles, as well as the therapsid proto-mammals. Even insects were hugely affected, the only known incidence in their entire history. It should be remembered that all current life on the planet comes from the scarce 4% that survived. The extinction that took place at the end of the Triassic finished 200 million years ago and involved the death of 80% of species; it was especially dramatic in the oceans, which saw the disappearance of the conodonts, almost all the marine reptiles and a large part of marine invertebrates, as well as many of the terrestrial reptile and amphibian lineages that had survived the previous episode. Finally, the Cretaceous extinction, which took place 65 million years ago and brought with it the death of nearly 76% of species, is one of the most well-known evolutionary events – as well as the most recent – and involved the disappearance of the dinosaurs and of the mosasaurs, pterosaurs and plesiosaurs, but also the final extinction of the ammonites and of other marine organisms.

Extinctions represent evolutionary opportunities for those who survive, who can occupy ecological niches that are left empty and diversify. Immediately after the Permian catastrophe, the lystrosaurus, who showed a mixture of reptilian and mammalian features, began to prevail in the fossil register and became the only vertebrate present on all the continents. In the same way, a grouping of small mammals similar to shrews survived the Cretaceous extinction and rapidly diversified into what were diverse lineages of mammals, among which primates are included. So, the disappearance of dinosaurs permitted the subsequent adaptive explosion of mammals.

The Sixth Extinction?

Nevertheless, it is possible that this will not be the last great extinction. The current environmental crisis has led numerous naturalists and paleobiologists to define a sixth extinction, which in good part coincides with the arrival of the warming period in the Holocene, which would begin with the disappearance of the megafauna – a category that encompasses large animals weighing more than 90-220 pounds, among them being, of course, the mammoths – and would continue with a wave of extinctions, above all of insular endemisms, as human beings expand and colonize the entire globe. From the beginning of the Holocene (that is, in the last 11,500 years), it is calculated that 177 large-size mammal species and 523 bird species have become extinct. It is interesting to take into account that human beings, in turn, also contribute to the creation of new varieties and species by means of transportation (around 891 invasive species have been recorded), domestication, hunting and the creation of new habitats.

What differentiates the last episode from previous ones is that it is influenced by, to a degree that is difficult to determine but is undoubtedly important, the presence of humans. For example, it has been ascertained that with the arrival of the Paleoamericans to America, the diversity of mammals fell from between 15 to 42%. In Australia, it is proven that humans arrived at least 50,000 years ago and rapidly expanded over the entire continent; the Australian megafauna, with paradigmatic animals such as the diprotodon subsequently, between 27,000 and 40,000 years, became extinct. Although it has been suggested that climate change, more than humans themselves, may have contributed to the extinction of megafauna in Australia, Eurasia and America, there is ecological evidence that the same evolutionary lineages had previously experienced other warming periods without any perceptible effect on their real demographics. Additionally, in many cases there is direct evidence (such as bones pierced by stone tools or having cut marks) that these animals were objects of active hunting. Everything indicates that human beings are the main suspects in these evolutionary crimes.

Climate change, a friend of extinction

Scientists agree that we are experiencing a peak in this new episode of mass extinction. It is calculated that in the last 500 years around 784 species have become extinct, which broken down comes to 79 mammals, 129 birds, 21 reptiles, 34 amphibians, 81 fish, 359 invertebrates (the majority mollusks), 86 plants and one protist. Additionally, the process is accelerating, and this year between 30 and 159 species – many of them even before being described – could be disappearing. These are alarming numbers that are added to the evidence of accelerating climate change, of which we are also responsible. It is fair to think that this dynamic paints a dark future for the tree of life on our planet in the short- and medium-term, and that in a few decades we will see the disappearance of corals, of the majority of amphibians and of large mammals such as the tiger, elephant and rhinoceros. The combination of human demographic pressure, global warming caused by the yearly release of seven billion tons of carbon dioxide into the atmosphere – with the resulting acidification of the seas – and the destruction of natural habitats could lead to a series of unpredictable cascading effects on the disappearance of species.

Knowing that there were past episodes of extinction, some could be tempted to argue that these form part of the natural cycle of life, that such events have later resulted in examples of new diversity and that, ultimately, they have led us to the present moment, where we are the dominant species. Nevertheless, conditions now are unique – even if only because they are marked by our capacity for planetary management as well as planetary destruction – and this new extinction could lead to the complete collapse of our ecosystems for millions of years. The economic losses and those in human lives alone could be dramatic and difficult to quantify.

The only good news is that everything indicates we are entering into this extinction and that a large part of its potential adverse effects could still be reversible. We have, as such, time to think again and change our impact on the natural world and to construct, as Edward O. Wilson extolled, a global ethic for the earth. And perhaps de-extinction will be one of the new mechanisms that help us manage this critical moment in our civilization and how it fits in the natural world.

The Road towards De-extinction

In the Gospel of John – the only one that has the following story and which, also, seems to be inserted in the wrong place as it breaks the narrative coherence – the story is told of the Raising of Lazarus. The sisters of Lazarus – one of whom was Mary, who had anointed Jesus with perfume and dried his feet with her hair – warn the prophet from Galilee that Lazarus is ill, but instead of immediately travelling to Bethany he takes his time, remaining where he is for two more days before going. When Jesus arrives, Lazarus has already been in his tomb for four days. The other sister, Martha, says to him: "Lord, if you had been here, my brother would not have died" (we can perceive here a somewhat reproachful tone). So Jesus replies: "I am the resurrection, and the life: he that believeth in me, though he were dead, shall never die." Taken to the tomb, he commands them to roll the stone away from the entrance and calls out: "Lazarus, come forth!" And as John recounts, "the dead man came out, tied hand and foot with cloths and his face wrapped in a shroud." We can imagine the astonishment of the crowd, if not a certain panic; as Borges said in *El Otro*, "those who witnessed the raising of Lazarus must have been horrified."

We have already seen how the evolution of life on earth is also synonymous with death and how, in the circle of life, extinction is a phenomenon that is as natural as the emergence of new lineages. A common characteristic in all death, whether individual like that of Lazarus or collective, is its irreversibility. A famous phrase attributed to Benjamin Franklin, but which we already encountered in Daniel Defoe, says there are only two things certain in life: death and taxes. But what would happen if extinction was not forever? What would happen if we could bring some of the species that have become extinct back to life?

For some decades, this idea has wandered around the margins of certain scientific disciplines without fitting into any one, and has been left parked in the realm of science fiction, with notable results such as the film directed by Steven Spielberg in 1993, *Jurassic Park*, and its sequels. Nevertheless, advances in genomic techniques over the past years have begun to show that de-extinction will be possible, and the controversy is not so much the "how" as the "when." In September 2016, for example, the Department of Ecology, Evolution and Marine Biology at the University of California Santa Barbara (UCSB) published a guide on which species should be selected so that their resurrection would have a beneficial effect for their respective ecosystems. And a recently created private foundation in San Francisco, "Revive & Restore," has the de-extinction of the passenger pigeon as its flagship project. But their objective is not only to revive extinct species to restore ecosystems, but also to improve current biodiversity, by means of genetic engineering, of species in danger of future extinction.

De-extinction seeks to resurrect and bring back to life forms that are similar to some of the extinct species from decades ago, or even from dozens of thousands of years ago. Although its defenders speak of "de-extinction" or "to resurrect" extinct species, this term can only be accepted as a metaphor. These species will never be brought back to life just as they were; for this reason, I think it is more correct to speak of "reinvention," as, whether this becomes reality or not, it will always be about genetic chimeras with components from current species. Even so, throughout this book and for questions of ease, we will use the three terms indiscriminately.

De-extinction, even without having seen any palpable final results, rests on four technological advances, the first three of which have taken place over the past 12 years: massive second-generation sequencing technologies, the emergence of synthetic biology, the CRISPR technique of genome editing and cloning techniques. The combination of these techniques, in conjunction with other more classic ones such as selective breeding, determines different possible approaches to de-extinction, which we will examine in detail in later chapters.

Criteria for De-extinction Candidates

As a first step to establishing the criteria for de-extinction candidates, we need to agree upon who is susceptible to being brought back to life. The researchers at "Revive & Restore" propose a list of three main criteria to decide if a species is, or is not, a good candidate for de-extinction:

1. Can an example of the extinct species be revived?
2. Can a viable population be restored from the extinct species?
3. Should this specific species be revived?

In the first criteria, we should take into account a basic question: restoration of all the member's genetic information, with a high quality of sequencing, must be technically possible. For this reason, it should be a species that has disappeared in the last thousand or dozens of thousands of years, but not much further back in time. With the dinosaurs, for example, there is no hope for a new life because their genetic material is not preserved. And it is not a technical question that can be resolved in the future. Additionally, we should have a reasonably close living relative so that some member can act as a gestational or surrogate mother. For those animals that represent unique lineages in the evolutionary tree, their return will be impossible. Even in the case of some who preserve living relatives, their biology may be so different that it impedes any attempt at de-extinction. For example, the South American giant ground sloths *(Megatherium americanum)* were, as their name indicates, enormous; they could reach nearly 20 feet in length and weighed between three and four tons. It is obvious that they were too big to be able to use a current sloth as a progenitor, given that these reach a maximum length of just under five feet. The same is true of the amazing glyptodon *(Glyptodon reticulatus)* who, although being a relative of current-day armadillos, was the size of a car and was covered in an enormous armored shell that gave it the look of a dinosaur. Although perhaps the most paradigmatic example of this problem is the litopterna *(Macrauchenia patachonica)*, a South American animal with an appearance similar to a camel although having three-toed extremities and a face with a short proboscis. It was just shy of ten feet long and reached nearly five feet in height at the withers; the most recent specimens date to only 11,000 years ago, but it is such a strange animal that an order (Litopterna) has been created specifically just for it. This means that no living relative exists that can be an obvious candidate as a surrogate mother and, in truth, it implies great difficulty to precisely map all its genome, given that in paleogenetics it is always necessary to have a genome of reference to distinguish the endogenous DNA sequences from the environmental ones.

In the second criteria, genetic and also ecological considerations come into play. To begin with, not only do we need to have genomic information from a member but we will need to have an idea of the diversity of the species. Cloning just one member over and over would be inviable from the genetic point of view, given that populations need a certain diversity to protect themselves from the negative effects of blood relationships and inbreeding. As such, it is necessary to create an entire population, stable over the long term and capable of self-support, and it is fair to ask if there are enough resources and the necessary amount of samples to obtain sufficient information on genetic diversity. An additional possibility would be to research mutations that are detrimental and which tend to accumulate, owing to inbreeding, in populations that have been small for a long period of time (which is almost always the case in species that have ended up extinct) and eliminate them from the de-extinct members by means of gene editing techniques.

But we should also consider if the species has a habitat it can return to or that could be restored, and if it has a sufficiently large expanse to insure its future survival if it is reintroduced. For example, the baiji, a dolphin from the Yangtze River *(Lipotes vexillifer)*, was declared functionally extinct in 2006 (the last known live specimen died in 2002, although a film in 2007 shows what could be another specimen in the water). But even if it was brought back from extinction and reintroduced into the river, the reasons for its disappearance (water contamination from frenetic industrial activity, as well as problems related to fluvial transport and hydroelectric dams) would continue to exist. It is calculated, for example, that nearly 530 billion cubic feet of waste water is released into this Chinese river each year, the majority without being treated in any way. That is to say, it makes no sense to dedicate efforts to de-extinct a species if the reasons for its extinction continue to be present; in fact, all signs indicate that, with the probable exception of deliberate fishing, the ecological degradation of the Yangtze will be worse in the future. On the other hand, in the case of the thylacine or Tasmanian tiger *(Thylacinus cynocephalus)*, a large part of the island continues being relatively unaltered and, from an ecological point of view, would allow sustenance for this species. It is calculated that in 2003, 95% of the 4.7 million acres of high-quality wild environment was found on natural reserves. In the case of the Xerces blue butterfly *(Glaucopsyche xerces)*, local to the San Francisco area and extinct in 1941 due to urban development, it has now been seen that different initiatives have restored the coastal dunes habitat where it lived and, as such, it would be possible to again maintain the species with the proper oversight.

Finally, another point to take into consideration is if the species in question is susceptible to being raised in captivity, even if only as a first step to its being set free in nature. As we will see in the chapter on domestication, there are many species that have difficulties to grow and reproduce in captivity, and this would be an added handicap in a project of this significance.

Many types of ecological, ethical, legal, cultural, economic and social considerations come into play in the third criteria, all of which must be explored and resolved. For example, would it make sense for us to bring back to life predatory species that are potentially dangerous to humans, such as the saber-toothed cat, cave lions, short-faced bears or the megalania – a giant Australian lizard that was a feared predator and reached a length of 16 to 23 feet and weighed over 830 pounds? It would also be necessary to investigate those species that could transmit diseases that might affect us. For example, there are diverse cases in which bats, owing to the characteristics of their immune system, act as vectors for diseases that affect us, and it would be logical that the de-extinction of a specific bat species took into account the possibility that it could turn into a vector for a new disease or one already known. But the opposite is also possible; some extinct species could have potentially beneficial effects for human health. There are numerous examples of substances obtained from the skin of amphibians that have been used successfully in the pharmaceutical industry, mainly against bacteria that are resistant to antibiotics. It has been speculated that Australia's gastric-brooding frog *(Rheobatrachus vitellinus)*, which became extinct in 1981 and incubated its eggs in its stomach (a period in which the frog inhibited the synthesis of its gastric juices), could be a source of new drugs to fight stomach ulcers. The invasive potential of the de-extinct species must also be taken into account; it could go from not having even one member to having millions of them, and its uncontrolled proliferation could end up affecting other species in the ecosystem. As such, any reintroduction must be perfectly controlled and monitored. Of course, this could be more of a problem with small-sized species who have a lower chance of being detected and whose recapture and monitoring is, as such, more problematic.

On the other hand, it is evident that de-extinct species would have an economic, cultural and social effect on nearby human populations in the areas where they are reintroduced, and the opinions of these people must always be taken into account. Undoubtedly, potential benefits would be created such as new opportunities in tourism, but there would also be space for complaints and injuries, in part because unforeseen negative effects could exist that might later lead to public opposition to a particular reintroduction, even though it was previously supported by the local community. For example, the Tasmanian tiger fed on sheep from the English colonists (which, in fact, led to them being hunted) and this activity continues to take place in Tasmania. Quotas would have to be established to compensate for each sheep killed by the reborn tiger. In any case, it is evident that de-extinction projects must be accompanied by exhaustive and detailed risk and benefit management plans, and should have the approval of local communities and international legislation. If the potential risks continue to be too large, the project should be aborted even if all other indicators, including restoration of the habitat and economic projections, were favorable.

The answers to these three important criteria will never be simple. Firstly, because we are walking in unknown territory, and later because a large amount of information must be taken into account and specific pros and cons must be evaluated for each case. No two species are the same, just like there are no identical ecosystems; each case will be unique and distinct from the others, and what is successful with one species could be a failure with another. Undoubtedly it will be necessary to involve scientists, ecologists, conservationists, politicians and even economists in each de-extinction project.

But independent of the, let's say, technical debate, it is evident that the arrival of de-extinction is a scientific and conceptual bombshell; that something as irreversible as death could in some way be reverted, affects our conscience and our imagination – that of scientists and of the public in general – and at the same time opens nearly endless scientific and ethical questions that we will attempt to address throughout this book. And let's not kid ourselves, everything that we can imagine scientifically can sooner or later be undertaken; and de-extinction is here to stay.

Chapter 1

De-Extinction Candidates: from the Dodo to the Woolly Mammoth

Given the high number of extinctions in the last 10,000 years, it is not difficult to find a list of de-extinction candidates, although these must necessarily meet a series of technical considerations, such as those set forth by the researchers at "Revive & Restore." For example, in a work published in *Trends in Ecology and Evolution* in 2014 from the Institute of Zoology in London, John Ewen and his collaborators determined, starting from similar criteria, a list of 20 firm candidates that are presented in the following table (the dates correspond with the death of the last known specimen, which normally precedes the "official" declaration of an extinct species).

Common name(s)	Scientific name	Region	Extinction
Passenger pigeon	*Ectopistes migratorius*	North America	1914
Carolina parakeet	*Conuropsis carolinensis*	Eastern USA	1918
Cuban red macaw	*Ara tricolor*	Cuba	1864
Ivory-billed woodpecker	*Campephilus principalis*	Southeastern USA	(1944)
O'o	*Moho nobilis*	Hawai'i	1934
Elephant bird	*Aepyorni sp/ Mullerornis sp*	Madagascar	(1800s)
Moa	*Dinornis spp.*	New Zealand	(1400s)
Huia	*Heteralocha acutirostris*	New Zealand	1907
Dodo	*Raphus cucullatus*	Mauritius	1662
Great auk	*Pinguinus impennis*	Northern Atlantic coasts	1852
Auroch	*Bos primigenius*	Europe, Asia, North Africa	1627
Pyrenean Ibex, Bucardo	*Capra pyrenaica pyrenaica*	Iberian Peninsula	2000
Thylacine, Tasmanian tiger	*Thylacinus cynocephalus*	Tasmania, Australia	1936
Woolly mammoth	*Mammuthus primigenius*	Northern steppes	(6400 yr before present)

Mastodon	*Mammut spp.*	North and Central America	(10 000 yr before present)
Saber-toothed cat	*Smilodon*	North America	(11 000 yr before present)
Steller's sea cow	*Hydrodamalis gigas*	North Pacific	1768
Caribbean monk seal	*Neomonachus tropicalis*	Caribbean	1952
Baiji, Chinese river dolphin	*Lipotes vexillifer*	Yangtze River, China	2006
Xerces blue butterfly	*Glaucopsyche xerces*	San Francisco, USA	1941

1. List of De-Extinction Candidates

Obviously, a variation in criteria and in evaluation of the effects of de-extinction (always debatable) would make this list vary, which in any case shows a relative lack of pre-Holocene candidates. This is because its greater antiquity negatively influences various aspects, such as having the same ecosystem available, recovering quality genetic information or in the difficulty of finding a viable surrogate female. We are going to briefly explain the history of some of the most paradigmatic cases.

The passenger pigeon *(Ectopistes migratorius)* is the flagship project for "Revive & Restore." Begun in 2012, its express objective is to make this bird de-extinct with the help of surrogate mothers from the band-tailed pigeon *(Patagioenas fasciata)*. The extinction of the passenger pigeon is a well-documented case; it is estimated that when Europeans arrived there were between three and five billion specimens in North America. It was probably the most abundant bird in the world. However, reduction in its habitat and uncontrolled hunting of enormous flocks caused its population to progressively decline between 1800 and 1870, and to drop dramatically in the following three decades. For example, in 1878 in Petoskey, Michigan, nearly 50,000 specimens were hunted with gigantic nets every day for five months. There was not time to design protective policies, because everyone remembered flocks with perhaps millions of members and it did not seem possible that it was in danger of disappearing. The last wild animal verified was hunted in 1900, and the last specimen died in captivity in 1914. It was not only an evolutionary tragedy: the passenger pigeon had huge ecological importance, as we will see, in maintaining the forest ecosystem in eastern North America. Its genome was sequenced in 2017.

The Carolina parakeet *(Conuropsis carolinensis)* is a similar case. It was the member of the parrot family that lived in the highest northern latitude of the planet, and was distributed from southern New England to the Gulf of Mexico, and all the way to the inner part of eastern Colorado. It had striking colors: green on the body, yellow on its head and orange on its face, and it formed noisy flocks of 200 to 300 members. In spite of its abundance, it began to decline during the nineteenth century due to indiscriminate hunting, and as of 1860 was only in restricted areas of Florida. The last wild specimen was hunted there in 1904, and the last specimen died in captivity at the Cincinnati Zoo in 1918. It is calculated that around 720 naturalized specimens are preserved, and around 16 skeletons. Recovery of mitochondrial DNA from a specimen has established that among its closest living relatives are parakeets of the *Aratinga* genus, typical of certain areas in South and Central America, who could be used as surrogate mothers. Some specimens show similar facial colorations, so it has been suggested that selective breeding could be carried out between these members. In January 2017, I took a sample from a preserved specimen at the Masjoan country house that had been acquired by the naturalist Marià Masferrer (1856-1923) at the beginning of the twentieth century. The sequencing of this sample at the Centre for GeoGenetics in Denmark has provided the first genomic draft of the Carolina parakeet, which we are currently analyzing.

The Cuban red macaw *(Ara tricolor)* was a parrot native to Cuba and the Isla de la Juventud that had an eye-catching combination of colors: its face, breast, abdomen and thighs were an intense orange; the tips of its wings and tail were blue, and its back and the beginning of its tail, brownish-red. It was hunted as a present for visiting dignitaries; this, in conjunction with the deforestation of its habitat, brought about its decline. The last known specimen was hunted in 1864 near the Zapata Swamp. In 2005, a new species of louse, named *Psittacobrosus bechsteini*, was discovered on the skin of a naturalized specimen of Cuban macaw and was apparently a parasite specific to this species. As such, this would be a case of co-extinction and constitutes an illustrative example of the complexity of the ecological interconnections involved in extinction processes.

The ivory-billed woodpecker *(Campephilus principalis)* was one of the largest woodpeckers in the world, given that it reached a length of over 1.6 feet. It lived in the forests in the southeastern part of the United States, but the disappearance of its habitat led to its probable extinction midway through the twentieth century. In spite of some possible – and controversial – sightings in 2006, and of generous rewards offered to those who could provide indisputable proof of their existence, there is no firm evidence that disproves it is already extinct.

The Hawai'i O'o *(Moho nobilis)* belongs to a family of extinct birds that were native to the Hawaiian Islands and are known generically as O'o. This specific species was described for the first time in 1786 and is characterized by its lustrous black color with some plumage of bright yellow. Europeans hunted them in large quantities for their beauty; in 1898 more than one thousand were still hunted, but soon they became scarce. The last confirmed sighting was on the slopes of Mauna Loa volcano in 1934.

Marco Polo explained in his notably imaginative chronicle of his travels that on an African island lived a bird named Roc, who was so big that he fed on elephants that he carried off by flying through the air. This story seems to come from a tale in *The Thousand and One Nights*, in which Sinbad the Sailor ends up flying away clutching onto the claws of a giant bird. Its eggs are so big that Sinbad ends up confusing them with the domes of a building. The probable origin of this eastern legend are the giant eggs that Arab and Indian merchants showed off when they returned from their trips to Madagascar. Everything is exaggerated, but it is indeed true that a bird lived on this island that was the heaviest that had ever existed: the elephant bird *(Aepyornis maximus)*. This bird belonged to the ratites group, which includes large birds, some extinct, who have lost their ability to fly, like the emu, ostrich, cassowary and moa. It is believed that the ancestors of the elephant bird evolved in isolation for around 70 million years after Madagascar separated from the African continent. The largest specimens were nearly ten feet tall and weighed half a ton. Their wings were very reduced, whereas the claws, which had to support their enormous weight, were very developed. Their eggs are an authentic Guinness record: they were 180 times bigger than one from a chicken and 15 times bigger than an ostrich egg. It has been calculated they had a capacity of nearly two gallons and that with just one you could make an omelet for 300 people. Elephant birds began their decline 2,000 years ago when humans arrived in Madagascar, transforming the landscape and no doubt hunting them.

When the first Portuguese explorers arrived, around 1500, there must have still been some specimens, and perhaps also in 1658, when the French governor Étienne de Flacourt described them (it is not clear that he saw one in person). In any case, around 1800 or perhaps a few decades later it was, with total certainty, extinct; a specimen was found with cut marks made by humans that had been carbon-14 dated to around 1,880 years. It's eggs are occasionally buried in the sand and are very coveted by collectors; it is known, for example, that the naturalist David Attenborough had one. Although in work done in collaboration with the paleogeneticist Alan Cooper in 2001 we were able to recover the first mitochondrial DNA sequences from the elephant bird, its phylogenetic position was not clarified until 2014, when in different work the same researcher recovered two complete mitochondrial genomes and verified that its closest living relative is the kiwi from New Zealand. This offers a possible surrogate mother, even if the difference in size represents a serious incompatibility (even so, the kiwi's eggs are notably big for its small body size).

Some islands, such as Madagascar or New Zealand, were rich in native animals when humans arrived. On the latter island, colonized by Maoris only 800 years ago, there was an entire grouping of large flightless birds known generically as moas. Due to the absence of mammals and large predators (with the exception of a giant eagle, the Haast's eagle), they did not need to flee and, as such, could lose the ability to fly. There were nine species of moas, that went from specimens the size of a duck to the *Dinornis giganteus*, which with its nearly ten feet of height and over 500 pounds of weight was probably the largest bird that existed. They had a strange, elongated appearance because, as opposed to other birds in the same group of ratites, they did not preserve any bone remnant from the upper extremities. Conversely, their claws were strong and developed. They laid enormous eggs, the size of a large soccer ball, having a capacity of 1.5 gallons. They were herbivores, and the different species of moas were adapted to exploit different eco-niches, from low shrubs to tree branches.

When the Maoris arrived, it is calculated that there must have been around 160,000 moas on the two main islands of New Zealand. These were hunted for food in such a way that by around 1445 they had become extinct. Although there were declarations of later sightings – some well into the twentieth century – it seems clear that upon the arrival of Captain Cook in 1769 they must no longer have existed. In 2001, I worked intensely with the researcher Alan Cooper on the first recovery of the complete mitochondrial genome of two species of moa. A project is currently underway to try to recover their nuclear genome with an eye towards possible de-extinction.

The huia *(Heteralocha acutirostris)* was a bird native to the northern island of New Zealand. It showed the greatest sexual dimorphism of all the birds in the world in the shape of its beak (to the point that initially it was thought that they were two distinct species): the male had a short and straight beak, while the female had an attractive slim beak, very long and downward curving. They were greenish-black with an orange patch at the base of the beak and were hunted to use the long tail feathers ornamentally, and also for museum collections. This, along with deforestation, led to their extinction at the beginning of the twentieth century. The last confirmed sighting was in 1907 in the Tararua mountains. An ongoing project currently exists to obtain its genome from museum specimens.

The dodo *(Raphus cucullatus)* is the symbol of recent extinction, but is also saddled with the stigma of being a reference to something obsolete and clumsy. In Victorian England, the saying "as dead as a dodo" became popular, and the bird itself was known for its appearance in *Alice in Wonderland* (it has been speculated that it is the alter ego of Lewis Carroll, whose real name was Charles Lutwidge Dodgson and whose stuttering often made him introduce himself as Do-do-dodgson). It was an extravagant-looking bird native to the island of Mauritius in the Indian Ocean. When the Portuguese came to the island, in 1598, they discovered these strange birds that did not fly and let themselves be picked up in someone's hands; it appears their name came from a Portuguese expression that meant something along the lines of idiot. They weighed between 44 and 50 pounds, measured nearly ten feet, had a beak that was enormous and curved like a hook, very short wings and some white feathers on their backside that gave them a rather comical appearance. European sailors who anchored in Mauritius stocked up on fresh meat hunting the inoffensive dodo by the hundreds; at the same time, dogs and pigs were introduced who ate their eggs, which they laid on the ground. By 1681 they were extinct. Few remains are preserved of the dodo in European museums, perhaps some 15 partial skeletons, although recent excavations in a marshy area of Mauritius have unearthed bones of new specimens. Genetic analysis of the remains found at the Museum of Natural History in Oxford – the only one having some soft tissue preserved, which is mentioned for the first time in 1656 – determined that it was a relative of current-day pigeons, which could be used as surrogate mothers. However, the difference in size could pose added difficulties.

Along with the dodo, the great auk or penguin of the Northern Hemisphere *(Pinguinus impennis)* is a symbol of recent extinction. These days there are only penguins in the Southern Hemisphere, even though a little more than 100 years ago there were similar birds – although not directly related – in the Northern

Hemisphere, and in large quantities. In fact, the name penguin originated with the great auk, and European sailors later used it for similar birds they found in Antarctica. The great auk was the only sea bird unable to fly in the entire Northern Hemisphere, where it occupied vast regions of the North Atlantic, which included Iceland, the Hebrides, the Faroe Islands, Newfoundland and Greenland. They were adapted to swimming and submerging themselves in water to fish, weighed around 11 pounds and measured some 2.3 feet; they had a black back – as well as a strong black beak – and a white belly. The great auk created large colonies with thousands of members and stood guard over only one cream-colored egg. Their inability to fly made them an easy target to be captured, and they were used to serve as supplies for the boats. The first voices of alarm were raised in 1785 as the colonies began to disappear, but when they became less frequent, European collectors began to pay substantial sums of money to get specimens. The last two specimens, a male and a female, were hunted on June 3, 1844, on the island of Eldey, near Iceland, by men paid by a private collector. The tally of specimens in distinct museums shows 81 naturalized specimens, 24 complete skeletons and 75 preserved eggs. The foundation "Revive & Restore" seeks to bring the great auk back to life using specimens from the razor-billed auk *(Alca torda)* as surrogate mothers. It has been proposed to place an initial colony on the Farne Islands in England, famous for their enormous colonies of sea birds. The great auk genome has been sequenced although is not yet published.

The auroch *(Bos primigenius)* was the wild ancestor to cows, which were domesticated in the Middle East around 10,000 years ago and entered Europe with the first farmers. They were stockier than their domesticated counterparts, and were characterized by having large horns. These wild relatives roamed on the continent in parts of Central Europe, where they were hunted for thousands of years. In the Middle Ages the scarce populations that remained were used for hunting by nobles and monarchs, until the last specimen died in the Jaktorów Forest (Poland) in 1627 of natural causes. Currently diverse genome projects are underway for selective breeding with current bovine varieties, as we will see further on.

The Pyrenean Ibex/Bucardo *(Capra pyrenaica pyrenaica)* was a subspecies of mountain goat restricted to the Pyrenean mountain range. Its horns were longer than the other three subspecies of mountain goat that have lived on the Iberian Peninsula, and in fact these were the principal reason for their indiscriminate hunting throughout the nineteenth century. The last specimen, a female named Celia, died of natural causes on January 6, 2000; an attempt at cloning derived from biopsy cells that had been frozen was later undertaken, which failed, as we will see further on in more detail.

The thylacine or Tasmanian tiger or Tasmanian wolf *(Thylacinus cynocephalus)* is another symbol of extinction brought about by human beings; it disappeared so recently that even films and photographs are preserved. In fact, the last specimen died alone in a cage at the Hobart Zoo on September 7, 1936. Later, even in this century, diverse cases of possible sightings have come to public attention, but none have provided incontrovertible evidence. It is called a wolf for its morphological similarity to the canidae, but in reality the thylacine, like all Australian mammals, was a marsupial. In the absence of placental mammals, thylacines diversified occupying all the ecological niches, which produced surprisingly similar evolutionary copies. The Tasmanian tiger occupied the highest position on the Tasmanian food chain. Its body brought to mind a large dog: it weighed from 33 to 66 pounds and measured around 2.6 feet in length, plus more than 1.5 feet of tail. Its head was disproportionately large, and it was famous for being able to open its mouth in an amazing bite of more than 120 degrees. The back part of its body was striped like a zebra, with around 13 to 21 dark stripes over the yellow-brown background of its body. This animal lived in the densest forest areas of the island, where it fed on small prey it hunted, preferably at night. The Dutch explorer Jacobszoon saw its tracks, which reminded him of a tiger's, in 1624. After the establishment of the English penal colony on Hobart, the flocks of sheep increased spectacularly and became the economic engine for the island. A black legend emerged that Tasmanian tigers attacked the flocks, and the colonists began to hunt them indiscriminately. As of 1888, the local government paid a reward for each animal killed (currently they offer substantial amounts precisely for the opposite reason: to anyone who can show their existence); in scarcely 21 years, they were able to quantify 2,268 dead specimens, a very high number for a carnivore. The Tasmanian tiger became increasingly uncommon, up to the capture of the last specimen in 1933.

2. Digital recreation of an example of a thylacine

Currently, a group from Pennsylvania State University directed by Stephan Schuster is trying to recover the complete genome from two thylacine specimens, from which he has already published the mitochondrial DNA. Nevertheless, an added difficulty is the fact that its closest living relatives, the numbat and the Tasmanian devil, are lineages that diverged from the Tasmanian tiger no less than 40 million years ago. Thus, to use them as surrogate mothers with a guarantee of success is, at the very least, doubtful.

Besides the famous woolly mammoth *(Mammuthus primigenius)*, which we will speak about in various chapters throughout the book, another representative of the proboscidea order, which had separated from the lineage of mammoths and current-day elephants around 24-28 million years ago, existed in America: the mastodon *(Mammut americanus)*. The mastodon weighed between four and five tons and was as tall as the mammoth; its tusks, however, were not curved like theirs but rather straight, and its cranium was much flatter. It seems to have preferred to live in conifer forests; as these began to diminish with the arrival of the Holocene, and in conjunction with hunting by the Paleoamericans, the mastodon population began to fragment until it disappeared. The first mastodon tooth, which weighed nearly 4.5 pounds, was discovered in 1705 in New York, and the animal to which it could have belonged was unknown (it was thought that the proboscidea were uniquely African). In the following years more bones were discovered, until it was described by French anatomist George Cuvier in 1806, who gave it its current name. The most recent remains are dated to approximately 10,500 years ago. In 2007, a team directed by the researcher Michael Hofreiter recovered the complete mitochondrial genome of a mastodon and was able to resolve its phylogeny. Three years later, the same team recovered a small part of its nuclear genome.

There are few names that capture the imagination quite like an animal that was "saber-toothed." Hidden behind this provocative name is a powerful carnivore, who surely terrorized human beings in the past. The most well-known is the *Smilodon fatalis*, which had the size and complexion of the present-day lion; it was strongly muscular and had a robust and compact body, probably designed to lie in ambush for prey rather than chase it over long distances. The trait of these carnivores that most stands out is, without a doubt, their spectacular upper canines, which reached a length of over seven inches and whose exact function remains a reason for debate among experts. It appears they would hunt large herbivores, but it is not clear if they would use their large incisors to pull their prey apart or to smother them clamping their powerful mandibles around their neck, as present-day lions do. Be that as it may, the saber-teeth seem to have emerged independently throughout the evolution of diverse lineages of these carnivores. Perhaps, as other researchers have suggested, they played a function to attract females during the courtship prior to mating. The *Smilodon* disappeared approximately 10,000 years ago in America with the arrival of humans to the continent, and in parallel with the disappearance of its potential prey, such as the mammoth and the mastodon. Fragments of its mitochondrial DNA have been published, which have been analyzed from a phylogenetic perspective. At this time its complete genome has been sequenced, although it has not yet been published.

When Vitus Bering, who discovered the straight separating Asia from America, found himself abandoned along with his men on the island that today bears his name, after being shipwrecked near the coast of Kamchatka during the winter of 1741, the expedition's doctor, Georg Wilhelm Steller, set out to study the nature surrounding him as he buried, one after the other, the explorers, including his captain. On July 12, 1742, Steller and his surviving companions hunted a strange sea animal that gave them – along with the seaweed they ate to avoid scurvy – the necessary energy to build a boat and sail to their salvation. According to Steller, the meat of the animal "had the flavor of ox and its fat tasted like almond butter." This was a member of the sirenian family (currently including manatees and dugongs) that is known as Steller's sea cow *(Hydrodamalis gigas)* and was adapted to the area's cold waters. It fed on marine plants, was between 23 and 26 feet long and weighed between 4.5 and 5.9 tons. Steller mentions that it did not have digits on its forelimbs and the catalog of specimens conserved in museums seems to confirm this strange anatomical detail (the best specimen is without a doubt the one in the Finnish Natural History Museum in Helsinki). Russian sailors set out to hunt it intensely in the following years, to the point that a mere 27 years after its discovery it had become extinct. It is thought that it was already a diminishing species, given the limited area of where it was located, and that possibly there were no more than 1,500 members when it was discovered. It is also possible that its disappearance was related to the intense hunting of seals in the area; their decline brought with it an excessive increase in sea urchins, who destroyed the marine plants that made up the sea cow's food source.

The Caribbean monk seal *(Neomonachus tropicalis)* was the only known seal living in tropical waters. Christopher Columbus mentions it on his second trip, in 1494, when a group of men killed eight specimens resting on the beach, south of the island of Hispaniola. It was originally found in the Greater and Lesser Antilles and along all the coastal areas of the Caribbean Sea. It had a length of between 7.2 to 7.8 feet and weighed around 286 pounds; its coat was chestnut, except on the belly where it was a brownish-yellow color. Before its decline, it is calculated that there were between 233,000 and 338,000 specimens. They were hunted massively, first as food for sailors and later to get oil for lamps and as a lubricant for the machinery on sugar cane plantations. At the same time, the exploitation and destruction of coral reefs restricted their habitat. Near the end of the nineteenth century they were already scarce; the last confirmed sighting dates to 1952, in Serranilla Bank, a group of small coral islands between Nicaragua and Jamaica.

The baiji *(Lipotes vexillifer)* was a freshwater dolphin found only in an area of some 1,700 kilometers of the Yangtze River in China, from Yichang to its outlet near Shanghai. The males measured nearly

7.5 feet and the females a bit more; they could weigh nearly 300 to over 500 pounds. They were a pale blue-grey color on their back and white on their belly. In comparison with their oceanic relatives they had very small eyes, and navigated in the murky waters of the Yangtze using an echolocation system (the dolphins in the Ganges and Indus have lost the crystalline of their eyes and are therefore blind). Increasing pollution in the river, associated with the Chinese industrial revolution, made their population decline drastically throughout the twentieth century until reaching their presumed extinction in 2006.

The Xerces blue butterfly *(Glaucopsyche xerces)* was a species native to the coastal sand dunes of the Sunset District in San Francisco. It was described for the first time in 1852, and was notable for its lovely blue color with white spots. The destruction of its restricted habitat from the city's urban development, as we have seen, brought about its extinction; it was seen for the last time in 1941 in the region of the Golden Gate National Recreation Area, where it occupied an area of only 495x226 square feet. It is one of the few insect species that appears among the bets on de-extinction. It depended on plants from the *Lupinus* and *Lotus* genus, which are still present in the area, and like other butterflies their larva has a symbiotic relationship with a species of ant that disappeared from the area due to an invasive species from Argentina; it is believed, however, that without these ants the larvae could grow. It is thought that it would not have any negative effects on people nor on other species, which makes it a strong candidate to be the first revived species. Its gregarious behavior in the months of March and April also imply that the process would be easy to revert if any danger was detected.

Candidates on the Reserve List

The species we have detailed are the ones that currently appear most frequently on different lists of possible candidates for de-extinction, and they are also the ones where there exist, to a greater or lesser extent of development, genomic projects already in progress and even projects of selective breeding or genome editing. Nevertheless, it is evident that the list could be expanded to other organisms, and that the discovery of well-preserved specimens could lead to some that are not here now being added to the lists. In this sense, only our imagination (and the paleontological record) is the limit. We could think, for example, about the woolly rhinoceros, Irish elk, giant sloth, glyptodon, giant capybara, bluebuck, the mouse-goat, litopterna, the cave bear or the Australian gastric-brooding frog.

In fact, the potential list is enormous: just on Réunion Island, in the Indian Ocean, 19 endemism became extinct between the seventeenth and eighteenth centuries, of which 11 are unique species of birds. And only on the islands of Martinique and Guadalupe, 17 endemism have disappeared in recent times, which include mammals, reptiles, mollusks and birds. In Australia, the last two centuries have seen the disappearance of the Darling Downs hopping-mouse (1840s); white-footed rabbit-rat (1840s); big-eared hopping-mouse (1843); broad-faced potoroo (1875); eastern hare-wallaby (1890); short-tailed hopping-mouse (1894); Alice Springs mouse (1895); long-tailed hopping-mouse (1901); pig-footed bandicoot (1920); red-necked wallaby (1927); desert bandicoot (1931); lesser bilby (1931); central hare-wallaby (1931); lesser stick-nest rat (1933); thylacine (1933) and the crescent nailtail wallaby (1964). As we can see, material is not lacking to continue down the road of de-extinction.

Chapter 2

Return of the Woolly Mammoth

The woolly mammoth *(Mammuthus primigenius)* is, without a doubt, the paradigmatic candidate for de-extinction, and the one that currently has the most genetic and functional information available. Understanding the advances in knowledge about its physiology and adaptive aspects will help us evaluate the technical difficulties of its possible de-extinction. Its majestic appearance and large size (they reached around 13 to 16.5 feet in height at the withers and weighed between six and eight tons) have made them the iconic animal from the Ice Age. For reasons difficult to objectify, we usually have an empathetic connection with them, just like we have with present-day elephants, highlighting their humanizing qualities such as intelligence, parental care or long-term memory. This does not happen with other animals that, in general, seem less sympathetic to us.

This fascination with mammoths is not new; we also intuitively sense this in the humans who coexisted with them – and who, almost certainly, contributed to their extinction. We can find representations of them in numerous cave paintings, such as Chauvet or Pech Merle, from the Upper Paleolithic period. In the Rouffignac cave (Black Périgord), for example, there are more than 150 representations of mammoths (which make up 70% of all the animals painted there and 30% of all the representations known in Europe). There are also utensils and figures carved in their ivory tusks. Perhaps the most impressive evidence is the Löwenmensch figurine from the Hohlenstein-Stadel cave in Germany, which was discovered in 1939 (some additional fragments were discovered between 1997 and 1998, which could fit in the original piece). The figure, carved from a mammoth tusk and dated to around 40,000 years ago, is almost one foot long and is probably the most ancient example of figurative art. At the same time, its clearly magical nature gives it an air of mystery: it shows the body of a man with the head of a cave lion. Mammoths not only provided physical support and imaginative substance for the humans' spiritual world in the Ice Age; these people also used their remains, which included bones, tusks and skin, as construction elements, funerary items and even as fuel.

In Ukraine, along the Dnieper River, different huts have been found built with hundreds of carefully   arranged mammoth bones. Similar structures have been found, some dated to 23,000 years ago and others from eras even later, in Poland and in the Czech Republic, and they are considered to be the most ancient examples of human architecture. The most famous hut comes from Mezhyrich, in the Ukrainian district of Kaniv Raion, which was discovered by accident in 1965 (today you can see a reconstruction in the Natural History Museum of Kiev); it is a circular structure around 33 feet in diameter with a fireplace in the center, constructed by a process of stacking hundreds of mandibles and large mammoth bones and using the tusks as hoists on which they perhaps hung skins and foliage.

In the case of Mezhyrich, it is calculated that they needed a minimum of 95 mammoths for the central structure, although their origin is not clear. It is not known if they were pieces hunted by the group or if they were remains that accumulated naturally and were collected (or a combination of both). In any case, the mammoths, with their own physical remnants, contributed to the survival of these humans who colonized the steppes and survived there under terrible weather conditions in the last Ice Age. They found shelter literally among their bones.

3. Mammoth specimens

It is probable that they also contributed with their meat to feed Paleolithic hunters, but it is not clear if they actively hunted them, nor how they did it. In spite of the numerous heroic representations of this sort in diverse museums and books, experts have not found precise details of this kind of hunting. Although the majority have no problem admitting that experienced hunters could camouflage themselves in the tall grass on the mammoth steppe and get close to these enormous beasts (who most likely, like present-day elephants, did not have very good eyesight), what might happen after is a contentious issue. If they actively attacked them, what weapons did they use? It appears difficult that they could penetrate the enormous fur covering and subcutaneous fatty tissue of an adult animal with their stone spearheads. In archeological iconography, traps are frequently depicted in the form of huge holes the animals fall into where the humans finish them off; but in the cold conditions of central and eastern Europe, where the ground would be permanently frozen, it would not be viable to dig such holes (even today, a powerful excavator would not be able to do it). Nor is there any firm evidence that they could ambush them and throw a huge stone on their head from a cliff. The accumulation of bones in La Cotte de St. Brelade, on a cliff on the island of Jersey, was interpreted as evidence of collective hunting, and that mammoths were made to fall off a cliff so their meat could then be used. But a recent reinterpretation of the site seems to suggest that they are specimens that, rather than being contemporary, were instead transported there and accumulated over many generations. In 2003, at the Lugovskoe site in Siberia, a mammoth vertebra was found with a Mousterian spearhead stuck in it, which indicates that hunting on the part of Neanderthals must have been a reality. But everything seems to indicate that it must have been an opportunistic and irregular activity that focused on attacking animals that were lagging behind, sick or young. In the Magdalenian period, around 12,000 years ago, we still find diverse representations of mammoths, such as in the Combarelles cave or the Madeleine (both in Dordogne), but later they stop being depicted. Be that as it may, it is evident that the final result of interactions between humans and mammoths was the extinction of these proboscideas.

The Phylogeny of the Mammoth

The ancestors of the mammoths left Africa and expanded through Eurasia nearly three million years ago; some populations in the areas of China and Siberia adapted to the cold conditions on the steppes and progressively supplanted other previous forms (the final evolutionary lineage of the mammoths includes diverse species which are, from the most ancient to the most recent, *M. meridionalis*, *M. trogontherii*, *M. columbi*, *M. jeffersonii* and *M. exilis* — these last three were only found in America – and *M. primigenius*). The forms adapted to arctic conditions had some characteristic features, such as smaller ears and tail – to preserve body heat – and dense fur with a thick layer of subcutaneous fat, as well as many sebaceous glands that helped repel water and improve thermal insulation. Around 200,000 years ago, the mammoth had colonized northern Siberia and later extended towards western Europe until arriving at the Iberian Peninsula. It also moved towards the east, crossed Beringia and entered the American continent; in this way, it occupied a large part of three distinct continents.

Nevertheless, the populations began to decline with the arrival of the last great Ice Age, some 20,000 years ago, and demographically it never recovered. Some isolated populations remained in northern Siberia and on Wrangel Island, where the last mammoths, considerably reduced in size and genetic diversity, became extinct only 4,300 years ago – that is, later than when the Egyptian pyramids were built.

From the beginning of ancient DNA research, at the end of the past century, the mammoth became an obvious objective for studies on the recovery of genetic material from the past. A clear scientific question existed that could indeed be answered with the limited technical possibilities of that era, which only allowed artisanal recovery of small mitochondrial DNA fragments: the phylogenetic relationship between mammoths and present-day elephants (African and Asian). Two studies published simultaneously in *Nature* in 1994 presented preliminary and contradictory results, which in part was due to the three proboscidean lineages having diverged in a short period of time, and to the fact that the females in this group of mammals did not disperse – as the males do – which tends to produce genealogies that are geographically very deep and structured when only mitochondrial DNA is looked at. In fact, these dispersion patterns differentiated by sex have produced conflicting genealogies between mitochondrial DNA and nuclear DNA in current African elephants (that is, between the forest elephant, *Loxodonta cyclotis*, and those from the savannah, *L. africana*). Today, with more genetic information available, it is estimated that they diverged between seven and five million years ago, firstly African elephants and, secondly, mammoths from Asian elephants, who would therefore be their closest living relatives. Although in that era this was perceived as being a purely taxonomic exercise, we now know that this is key information for de-extinction, which requires knowing precisely the living species that is, from a genetic perspective, the most similar to the one that is sought to reinvent.

The Genetic Diversity of Mammoths

The following steps in genetic exploration of mammoths focused on another objective that, without knowing it then, was also crucial in de-extinction: knowledge about the species' global genetic diversity. Diverse studies, based on mitochondrial DNA, determined the existence of three large clades of diversity (a clade is a grouping that contains a common ancestor and all the descendants – living or extinct – of this ancestor) throughout the vast distribution range of mammoths. These studies culminated in a piece of work, which was published at the beginning of 2017 and in which I participated, where complete mitochondrial genomes were recovered from no less than 143 mammoths, including a wide, previously unknown region, Europe (among them, two from La Aldehuela, in Madrid, excavated in 1968). The study confirmed the existence of three large groupings of mitochondrial lineages: one on the American continent, another in eastern Eurasia and Alaska, and another in western Eurasia. This last one extended from west of China and Siberia to Germany, the North Sea (from where they had dredged up diverse specimens), and was even present in two specimens discovered in Madrid.

Global analysis of all these samples confirmed that the strong geographical structure of the mitochondrial DNA did not correspond with the morphological allocation from the distinct species of mammoth, nor with the genomic information (for the samples in which this was available). Once again, this confirmed the philopatric behavior of the females and the dispersion of the males who, just like present-day elephants, abandoned the group they were born into when reaching sexual maturity; as such, while mitochondrial lineages tended to remain structured in space, the populations harmonized on the genomic level thanks to the greater mobility of the males. This included crossbreeding between species, the same as had been discovered some years ago in hominins. At any rate, the study lay the foundations to try to include all the diversity of mammoths; in the case of wanting to reinvent them, it would have to be taken into account that they were mammals who occupied all the Northern Hemisphere of the planet, and without a doubt they would have, in addition to fixed genetic changes, other changes that would be exclusive to one region or another.

The Pigmentation of the Mammoth

The first symbolic step towards understanding the adaptive differences of the mammoth happened in 2006, in a study I participated in that was published in *Science*. Some months earlier, my friend Michael Hofreiter, an expert in ancient DNA who at the time worked at the Max Planck Institute in Leipzig, had sent a bone fragment from a Siberian mammoth dated to 43,000 years ago to my laboratory. He had recovered in 23 overlapping fragments the complete *MC1R* gene; this gene plays a leading role in the pigmentation of hair in mammals. For example, diverse mutations in human *MC1R* determine that some individuals are redheaded, as these mutations affect the role of the resulting protein (also called *MC1R*) in the membrane of the melanocytes and the way in which this protein interacts with the MSH hormone. The correct interaction between *MC1R* and MSH sets off the synthesis of a dark-brown pigment (called eumelanin) and in humans also contributes to the ability of their skin to turn brown when laying in the sun. Conversely, a dysfunctional interaction leads to the synthesis of an alternative pigment of reddish-yellow color, called pheomelanin, and secondarily, to a low ability to turn brown. This is the predominant pigment in individuals with reddish hair, and with all probability was also present in some Neanderthals, who showed similar (but not identical) mutations to those of present-day redheads. Hofreiter had analyzed four mammoth bones and found that two of them had what appeared to be three mutations in the *MC1R* gene. One of them in particular seemed to have a strong effect on construction of the protein and, as such, could be presumed to have a functional, although in practical results, unknown role.

Hofreiter asked me if I could replicate the results independently in my laboratory and that is what I did, after some weeks of work in the laboratory (at that time, this was the most secure way to confirm surprising results. Additionally, we were able to discover that the three mutations were present in a copy of the chromosome, but not in the other, which is known as heterozygous in genetic language). If the mutations acted like the ones in human redheads, they would need to be present in both copies to affect the external characteristic; that is, the color of the animals' fur. But once these mutations existed in the population, this was not a problem; the crossing of two heterozygote mammoths like the two we had found would provide, in 25% of the cases, members with the two copies of the mutations. That is, they must have existed. But what exactly would happen with these members?

Up to that time, all the work in paleogenetics had worked on understanding the sequence level, but none had tried to go further. This was due to an unfavorable combination of two factors; in the first place, very few studies had been able to successfully recover a nuclear gene and, secondly, the role of heredity in the majority of external characteristics is very complex and depends on numerous genes. There are few characteristics, like that of skin pigmentation, that in essence depend on only one gene. We tried to go further and explore the palpable meaning of what we had discovered. To do this, we carried out the first study on functional paleogenomics.

What we did was insert the two versions of the mammoth *MC1R* gene (with and without the encountered mutations) into pigment cells on petri dishes. When the gene was added, the cell membranes showed the mammoth protein, and later we had them interact *in vitro* with the MSH hormone, and we were able to measure in the laboratory the result of this interaction. We found that the mammoth haploid with the three mutations showed a 65% drop in interactive capacity between the *MC1R* and MSH (this phenomenon's larger proportion was due to a unique mutation, the third, which changed an amino acid, arginine, for another, cysteine, in position 67). The result of this drop was the synthesis of pheomelanin instead of eumelanin. As such, this showed us that there must have been mammoths with light or reddish fur, as well as with dark brown. We did not know, of course, if this characteristic had an adaptive role and, thus, if it was important to understanding mammoths, and we also did not know its frequency because we had only analyzed four specimens. Numerous mammals exist whose species or forms show different pigmentation (lighter or even white) in arctic climates. But normally it is a question of camouflage against a snowy background, and it did not seem that it would have been necessary for mammoths to go unnoticed (nor would they have had a way of doing so).

In any case, our study established the foundations to begin thinking that the external characteristics of mammoths and, in general, of extinct species should be able to be researched from the genes, and that knowing the first ones would help in understanding the others. We did not know it, but we had inaugurated the conceptual road towards de-extinction by means of the manipulation of informational genes. It is also evident that pigmentation is one of the key characteristics in the composition of a species, and that any future study should make this a priority if it wanted the de-extinct species, whatever it was, to have the "correct" appearance.

Our study had a curious finish. In 2010, in work published in which the *MC1R* gene was analyzed in no less than 47 mammoths, the majority from northern Asia and Alaska, only one other specimen was found with the three mutations and also in heterozygosis. With the existing sample size, this means the mutations of light fur were only present in 3% of the animals, a number that situated the characteristic in the realm of being anecdotal. It seemed clear that the majority of mammoths must have been dark, although as we will see later, the matter of pigmentation was not closed.

The Mammoth Genome

When we did our study, the technology only permitted the recovery of small fragments of DNA. But the parallel development in 2005 of new massive sequencing technologies allowed attaining complete genomes from extinct species to be addressed, which saw the ancient DNA discipline go from being a mere scientific anecdote to being one of the most dynamic fields in the study of the past. The first step toward the field of paleogenomics was taken by the mammoth itself.

In Search of the Genome

In January 2006, the magazine Science published the sequencing of 28 million DNA nucleotides from a Siberian mammoth bone dated to about 27,700 years ago coming from Lake Taymyr, and from which almost half, 13 million, were 98.55% similar to the DNA from the African elephant. It should be mentioned that, at the time, the elephant genome was not even completed;

in this one simple fact, the dynamism of one scientific field in respect to another that, on paper, has many less technical difficulties can be seen. The rest of the sequences aligned with fungi and bacteria, and 1.4% with the human genome (with all probability, theses came from contamination by those who had discovered and handled the bone). Also, an additional and significant fraction, 9%, were most probably from the mammoth itself, but they could not be precisely situated in the still incomplete elephant genome. The authors concluded that, with these types of samples, it would soon be possible to obtain a complete mammoth genome in the future and interpret it from a functional perspective to understand the genetic basis of its adaptive characteristics. They were not wrong in their predictions.

The complete mammoth genome was published two years later in *Nature*, in work directed by Stephan C. Schuster from Pennsylvania State University. The authors obtained 4.17 billion nucleotides of sequence – which represented approximately 70% of the genome – from three distinct specimens, although the majority of the data came from a Siberian mammoth dated to 18,500 years ago. But this was a descriptive study that continued without delving deeply into functional aspects of the genome; that is, in the changes in certain genes that made a mammoth a mammoth and an elephant an elephant, in spite of sharing a large part of the same DNA sequence. Additionally, the quality of the genome was low, given that not all the positions were represented, and also the low number of sequences that covered each genome position implied that many of the genetic changes detected could have been sequence errors. These errors are produced by chemical processes that degrade the original DNA, and can be detected in ancient genomes with a high coverage of sequences, given that they are infrequent and produced stochastically. But a quality genome is an essential condition in any de-extinction task; we have to be sure that the changes detected are real and not mistakes. It should be remembered that just one genetic change that is not real introduced by mistake in the organism you want to reinvent could have catastrophic consequences in its viability, given that, by definition, new mutations often have damaging effects (little wonder they are produced by chance in a viable and functional organism whose genetic information is perfectly integrated).

The Mammoth's Adaptations to the Cold

After a couple of years, *Nature Genetics* published work directed by Alan Cooper that provided the first physiological evidence of adaptation to the cold as a characteristic of the mammoths. This was about the recovery, first genetic and later *in vivo*, of hemoglobin from the mammoth. These types of studies are the ones that will permit reinvention of a species in the future, because they are exactly the characteristics that interest us to take into account.

Hemoglobin is a protein from blood that takes care of transporting oxygen and freeing it in the cells, as well as collecting carbon dioxide and carrying it from the tissues to the respiratory organs. Nevertheless, as the temperature goes down, the affinity of hemoglobin for oxygen increases, which damages its capacity to efficiently free oxygen from the tissues. It was not understood how mammoths, who lived in a frigid climate where the average temperatures in winter could be between 30 and 50 degrees below zero, could continue exchanging the vital oxygen for tissues. Researchers were able to recover the protein sequence from a 43,000-year-old bone and later synthesize it in the laboratory inside the *Escherichia coli* bacteria, where they could measure oxygen levels according to the different shapes of the hemoglobin. There they discovered that its unique structure provided a specific biochemical answer to the problem of arctic temperatures, and that hemoglobin in mammoths was designed to keep functioning correctly at low temperatures. This work marked a turning point: it is evident that if someone wants to revive something similar to a mammoth in the future, this animal needs to be capable of staying alive in low temperatures and its hemoglobin must be like this, and not like that of an elephant. However, it was only a characteristic, although key, in the functional comprehension of the mammoth, and it was not known how many more, one by one, would have to be characterized in the future.

In this sense, the most detailed work to date exploring the underlying genomic changes to adaptive aspects of the mammoth was published in 2015 in *Cell Reports*, and was again directed by Schuster. The researchers sequenced three Asian elephants and two mammoths of some 20,000 and 60,000 years, respectively, up to a genomic coverage of 20x (that is, each genome position was represented on average by 20 different sequences, which corresponds to quality that is similar to what is obtained in a standard way for a present-day human sample). They discovered that there were 1.4 million nucleotide changes in the mammoth lineage as compared with that of elephants, of which 2,020 represented amino acid changes that affected 1,642 proteins (there are nearly 20,000 proteins in the genome of mammals); they also located 26 proteins with some mutation that inactivated them in mammoths in respect to their proboscidean relatives (simply, they did not need these to live and many of them seemed to be associated with cholesterol and triglycerides). With knowledge on the function of genes, obtained over decades of experimentation on mice and in thousands of functional studies on humans, they could classify all these genetic changes in various categories:

changes associated with the unique morphology of mammoths, changes associated with circadian rhythms, changes related to the metabolism of lipids and the accumulation of fat, and changes involved in thermal adaptation. We are going to review some of these, as they are essential to understanding how a mammoth worked on the inside.

In respect to the first category, they found genes that had been described in other mammals as being involved in anomalous morphologies in the tail or ears. Frequently we do not know the exact function of genes, but only their biomedical consequences when said genes do not work well; in this case, it seems evident that they must be genes involved in making a proboscidea with small ears and tail. They also located changes in genes that had phenotypic effects on the cranium (it must be remembered that mammoths were characterized by having a globe-shaped, very high cranium and expanded parietals in the form of a ridge). Likewise, they found genetic changes in three genes that produced larger sebaceous glands. The subject of pigmentation reappeared, due to amino acid changes in 38 genes which, in the mouse and other mammals, had been described as being associated with changes in fur color. Of these, eight genes were related with light colors in other animals; the authors suggested that there was probably a variation in color that was greater than what had been determined in previous studies, based only on mutations in the *MC1R* gene. It was evident to them that there must have been mammoths with a wide range of colors, from blonde and ginger to dark-brown. And they also described another group of genes that were related to the morphology of hair follicles and hair stems.

In terms of circadian rhythms, it is evident that this is a powerfully adaptive matter for arctic species. Given that the sun does not set during the entire day in summer, and that the darkness can be complete in winter, these species cannot base their life cycles on the duration of sunlight or on daily changes in temperature. They often show specific mutations in genes related to circadian rhythms, such as *PER2* and *BMAL1*. The mammoth shows amino acid changes in eight genes related to the biology of circadian rhythms, such as *PER2*, *HRH1*, *HRH3* and *UCP1*. In mice, the elimination of these genes leads to the existence of anomalous rhythms, which include aspects such as going from eating in the day to eating at night.

As to the third category, another arctic adaptation of mammoths was described that consists in the transformation of 54 genes involved with metabolism, and which includes the synthesis and accumulation of lipids, development of adipose tissue and diverse physiological aspects. The existing information, above all from mice that have had these genes manipulated, was the appearance of phenotypes with anomalies in the shape and subcutaneous fat deposits. Among the modified genes in the mammoth were some previously known in human metabolic disorders, such as the Growth Hormone Receptor (*GHR*) or Leptin Receptor (*LEPR*), these latter ones involved in the differentiation of adipocytes. They also localized 39 genes involved in the functioning of insulin, which included some that had been studied in mice and whose elimination brought on abnormally high levels of this hormone or problems with glucose tolerance.

Finally, regarding the fourth category, the authors detailed an additional list of 13 genes that had attracted their attention, and which seemed to be related to thermal adaptation. Some of them, such as *TRPM8*, were involved in the perception of harmful cold, and others, such as *TRPV3* and *TRPV4*, in the perception of harmless heat. But it was known that others, like *TRPM4*, were encoded for proteins sensitive to changes in temperature, but did not seem to be involved in thermal functions. The researchers modeled some of these proteins with the mammoth mutations on a computer to understand how they affected their function in responding to cold conditions. A large part of their efforts focused on understanding what the structural consequences were from a protein, TRPV3, that was expressed on epidermal keratinocytes and whose function was known to be the sensation of heat (approximately more than 33°C) and also hair growth and the accumulation of fatty tissue. A change of amino acid in this protein in mammoths caused it to be less active in a wide range of temperatures, which would suggest a greater tolerance to cold and also, probably, a tendency to develop longer hair. For the authors, it was a mutation of singular importance in the adaptation of mammoths, due to the ripple effect in diverse aspects of its biology.

Incidentally, a recent functional analysis of the genome of a mammoth from Wrangel detected an impressive number of accumulated harmful mutations (among them, many that interrupted the synthesis of proteins), without a doubt a consequence of the low demographic efficiency of this rare isolated population. Additionally, two mutations had completely inactivated the *FOXQ1* gene; if the effect is the same as when inactive in mice, these mammoths must have had lovely satiny hair that would make them similar to a cuddly toy with a trunk. Evidently, the Wrangel genome would not work as a model for de-extinction of the mammoth, because we would create an individual with additional genetic problems from a long evolutionary decline.

Precisely identifying genetic changes that are related to adaptations of animals is a complex scientific field, especially when work is done with non-model organisms (the model mammal par excellence is the mouse) and with extinct species. But the list of genetic changes provided by the mammoth genome is the necessary initial condition for de-extinction. Not all these changes are going to be critical, and it is possible that some are absolutely unnecessary (that is, there can be changes to a gene that result in a protein whose function is, in practice, identical), but at least we now know what the shopping list is. Beginning now, scientists need to develop functional studies to start discovering which changes in the mammoth's more than 1,600 proteins are going to be decisive. It is arduous work, but not impossible, and from a scientific perspective perfectly justifiable, with or without the final purpose of de-extinction. But as to paleogenomics, once a quality genome and an idea of the genetic diversity of the species studied is achieved, the road to follow is no longer in its field. We, the paleogeneticists, recover extinct genomes, but it is not under our jurisdiction to recreate them. This will be the playing ground of a new scientific area: that of synthetic biology.

Chapter 3
Experimenting with Species: Domestication

During the 1950s, when Stalin began hunting down all the scientists who followed the ideas of Darwin and Mendel, an obscure researcher from Novosibirsk, Dmitri K. Belyaev (1917-1985), decided to use his scarce research funds on the task of exploring – undercover – the phenomenon of domestication using a program of selective breeding of the red fox *(Vulpes vulpes)*. His strategy consisted in allowing the reproduction, generation after generation, of members who showed less fear in the presence of humans; those who were discarded went directly to the Siberian fur factories. With patience, without resources and nearly anonymously (for the

Soviet authorities, he was simply trying to improve production of higher-quality fox furs), Belyaev carried out for several decades what has been described as "the most extraordinary experiment in breeding and reproduction ever done."

How to Domesticate a Fox

The consequences of the Soviet program form a cornerstone in the understanding of the processes by which humans experimented with the modification of species in the past, that is, domestication. Belyaev's final objective was to create foxes that were comparable to other species domesticated by humans during the Neolithic, that is, cows, sheep and pigs. Although he died in 1985, the project remains alive thanks to his disciple, Lyudmila Trut. The main premise was to research those species whose domestication failed, or was not carried out, and try to understand why this happened; for Belyaev, if selection was made for one essential characteristic, meekness, the entire process followed and was, in a certain way, unstoppable. At the beginning of the experiment, nearly 10% of the foxes showed the desirable characteristic of docility, and these were the ones selected to form the following generation; in the first step, 100 males and 30 females were selected to make up the new generation. The new foxes were fed by hand and, once again, those that showed less aggressive responses towards humans were selected.

As a response to this strategy, in two or three generations of foxes, the more fearful or violent members had been eliminated, and in only one more generation the researchers began to observe secondary changes – not associated with personality, which was the only one selected – that amazed them; the new pups began to behave like dogs, friendly and playful with their masters: they were able to "read" human gestures and emotions, had floppy ears, barked and moved their (hanging) tail when they were happy. Some of these characteristics were typical of young members, but the difference was that they remained in adult life. What surprised the international researchers who visited the center after the fall of the Soviet Union was not so much the scope of the evolutionary changes achieved, but their speed. For Brian Hare, a professor from Duke University in Durham, North Carolina, "the fact that in 50 generations [the foxes] were wagging their tail and barking is truly incredible." And what was even more remarkable, "in only a few generations, the friendly foxes showed changes in the color of their fur." According to Trut, in more advanced studies in the experiment, the foxes showed skeletal changes that included shorter paws and tail, as well as a slimmer snout and more delicate teeth.

4. Domesticated fox playing with its owner

By then, even their reproductive habits had changed; the new foxes could reproduce during longer periods of time and the "domesticated" mothers had on average more descendants. The Russian experiment was still alive in 2016, with nearly 270 females and 70 males from these domesticated foxes, but the dangers are again economic. For the current head of the center, Kharlamova, "the current situation is not catastrophic, but at the same time it is not stable." Since 1990, the Siberian institute has tried to support itself by selling foxes as household pets. Currently, a North American company, the Lester Kalmanson Agency Inc., imports foxes to the United States as pets for the modest price of around $8,900 per specimen.

What is most interesting about the Russian fox experiment is the confirmation that humans already have a mechanism to experiment with the nature of species (animals and also vegetables), and that this continues to be functional. Any ethical consideration we wish to make about the future possibility to modify the genetic patrimony of species should take into consideration what happened in the Neolithic and led to the domestication of dozens of wild animals. We are not new arrivals to the territory of modifying the genetic information of animal species; if anything, starting now these processes will be carried out in a more conscious and planned manner.

Domestication Candidates

In his famous book *Guns, Germs and Steel*, Jared Diamond explores when, why and how animal species were domesticated, and what were the consequences of these processes on the human societies who experienced them. For Diamond, domesticated species are a percentage of the candidate species from a continent, given that they meet a series of requirements; for example, in Eurasia there are 72 possible candidates, crystallized in 13 (18%) that are domesticated; in Sub-Saharan Africa, there are 51 candidates, but all of them are unsuccessful. In America, there is only one domesticated animal for around 24 candidates. Diamond tried to show that the later technological dominance of Europe could, in part, be attributed to this initial success, which was to a large extent contingent (for example, the horse was able to be domesticated on the Eurasian plains, but due to a question of nature, its African relatives, the zebras, never allowed this).

That is, there was nothing intrinsically superior or good in the Europeans, all our later fate was a question of chance related to the successful roulette wheel of domestication. Independently of this argument, which has been somewhat controversial, Diamond provides a series of characteristics that domesticated species seem to meet, especially those that he calls the "Major Five" (and to which should be added the dog, without a doubt the first animal domesticated by humans). These are the sheep, goat, cow, pig and horse. After these come the "Minor Nine," named as such because they are more geographically restricted; these are the Arabian camel (with one hump), Bactrian camel (with two humps), the llama and alpaca, the donkey, reindeer, water buffalo, yak, Bali cow and the mithan (whose wild ancestor would be the gaur). To this list should be added the cat, the so-called "domestication failure" – for its independent nature – and some later cases, such as rabbits, or smaller animals like chickens or pigeons.

The majority of the "Major Five" ancestors were found in Eurasia (their exact origin is difficult to discern, due to multiple hybridization events with wild ancestors). The geographical layout of the continents (whose orientation, in the case of Eurasia, is basically a longitudinal axis) favored their later dispersion, in contrast to continents that, like America and Africa, had essentially latitudinal axes and, as such, were more susceptible to having ecological and climatic barriers over long dispersion routes.

The domestication of all these animals, which dates to some 10,000 years ago, begins in a similar way and with purely contingent methods: it is about choosing for reproduction, generation after generation, those specimens that meet the most favorable conditions. Although at first it was about selecting behavioral characteristics that were friendly towards humans, in later processes they sought to favor specific characteristics, such as greater milk production, more wool or more meat. It is not surprising that these domesticated animals have thinner legs and less corpulent bodies than their wild counterparts (because they require less mobility), and also smaller brains, given that the cerebral areas needed for reaction and flight mechanisms stop being necessary. There are also behavioral mechanisms involved in the candidacies; usually domesticated species are those that have strong hierarchical structures where humans supplant the leader of the pack. This social change happened in the case of wolves, horses, sheep, goats and cows. Additionally, a mark is made on the new offspring from these domesticated flocks by the humans who herd them. In contrast, territorial or solitary animals are the ones that have failed the domestication test. As social structure is in part related to diet, the majority of domesticated animals are herbivores (dogs and cats are an exception, and next are pigs, omnivorous animals).

Even taking into account all these factors, Diamond shows that, of the 148 medium-to-large-sized herbivores that were candidates for domestication, only 14 of these passed the test. The reasons for failure of the other 134 are varied. In some cases, the growth rate is very slow and makes raising some of these species, such as elephants or gorillas, economically inviable. In other cases, it has to do with species that do not accept being raised in captivity, at times because their reproductive strategy requires complicated mating courtships that are inhibited in captivity (as in the case of llamas, for example). Other reasons are related to temperamental characteristics, which turn some candidates into animals that are unpredictable and dangerous for humans. This is the case of bears and also zebras; the latter are sadly famous in zoos throughout the world for their tendency to bite the fingers of their caretakers (and to then not open their mouth under any circumstances). Finally, some species have panic attacks when they are fenced in and tend to jump to try to escape, including at the cost of repeatedly smashing themselves against a wall. This is the reason why it has never been possible to domesticate antelopes like the gazelle.

The Genetic Transformation of Domesticated Animals

The modification that humans subjected domesticated species to was so large and generalized that, in some cases, like in horses or cows, the wild ancestor completely disappeared (it has recently been seen that what is known as the Przewalski horse is not a direct ancestor of current horses). In other cases, such as with pigs and boars, or dogs and wolves, these ancestors lingered in vast populations and even repeatedly hybridized with their domestic relatives. These events could happen inadvertently when flocks were grazing in open fields, or even intentionally to favor adaptation of the domesticated animals to local ecological conditions.

But in all the cases, the enormous disparity between the original forms and the descendants is clearly seen. In some vegetables (such as corn), the original plant, teosinte, is so different that it is barely recognizable as the native precursor. The changes are evident physically, but also genetically. For example, in a study on genomes from dogs and wolves published in *Nature* in 2013, it was verified that the first differed from the latter in multiple genes that had been selected during the domestication process, and which included, for example, changes in the *SGLT1* gene, involved in the transport of glucose in the blood. Researchers deduced that dogs had adapted to ingesting derivatives from starches, predominant in farmers' diets, and had as such strayed from the wolves' typical carnivorous diet. They also detected changes in behavioral genes (such as *GRIK2*, *GRIK3*, *MECP2*, *GABRA5* and *BCL2*), which made dogs feel less fear and anxiety and, as such, be able to approach humans at closer distances.

It has been suggested that this was probably the initial driving force behind their domestication, thousands of years before the appearance of agriculture (and, as such, a diet rich in starches); it is believed that the first proto-dogs drew near looking for food that groups of Paleolithic hunter-gatherers had cast aside. Canine ancestors that had these variants had more possibilities to survive, and as such they ended up prevailing. Other selective drivers, like the ability to correctly "read" human emotions or gestures (something current wolves are not very good at), also imposed themselves on these first selective baby steps. Similar changes have been described in behavioral genes in other domesticated animals, such as the horse. In this case, they have also found additional changes in 125 genes that seem to be associated with the development of muscles, joints and extremities, and could be related to mounting on the part of humans (the case of the horse is an unique example of physical integration between the domesticated animal and the person who domesticated it, the person who rode it; if we really think about it, it is truly incredible).

With the probable exception of the dog, all the remaining domestication processes took place during the Neolithic, and date to around 10,000 to 4,000 years ago. After what was seen on the fox farm, there is no doubt that there is more than enough time to create the notable genetic and morphological changes that we see today. Even we have not escaped this process; some mutations selected in recent times in Europeans, such as the ability to digest milk as an adult (coming from a mutation near the *LCT* gene) are not more than a few thousand years old and, in a certain way, can be considered a secondary product of domestication, which in this case affects its instigators. It cannot be ruled out that we can also find changes in behavioral genes, perhaps associated with the possibility of living in large hierarchical social groups, which can also respond to this post-Neolithic social creation process.

A common characteristic in many domesticated animals is the serious loss of genetic diversity as compared with that of their wild ancestors. This is because the domestication process, which is often unique and has only one initial focus, results in a strong demographic bottleneck. The loss of diversity, which is also seen in species in danger of extinction, where few live specimens of a species remain, involves the appearance of negative effects as a secondary aspect due to the accumulation of harmful genetic variants. This phenomenon is known as "the price of domestication," and can be dramatically seen in recent varieties of cats and dogs that, in some cases, have been created from very few initial individuals and suffer from high incidences of certain disorders.

Domestication – a Lethal Gift

In spite of the benefits for humans brought by domestication, it did not take place without paying a high price: given the physical proximity we had to share with these animals, it is not surprising that their microbes also invaded us. The majority of infectious diseases that afflicted us and still afflict us, such as smallpox, flu, measles, chicken pox, brucellosis, tuberculosis, rubella, mumps, etc., appear to come from domesticated animals. Although rarely lethal today, they undoubtedly represented big evolutionary challenges for our farming ancestors, whose genome was by no means ready to fight them. It is necessary to remember that we are in reality descendants of those who survived; but many fell by the wayside, owing to this "lethal gift," as Diamond called it, from domestication.

You need only look at what happened when Europeans arrived in America, a continent that had been isolated from these contacts: illnesses that were already benign for Europeans brought about huge death tolls in the Amerindian populations (in some cases, it is calculated that a 90% mortality rate was reached, and without a doubt they contributed to the rapid collapse of some civilizations on this continent). Zoonosis, the transmission of pathogens from domesticated animals to human beings, was the price early farmers paid for playing at being gods and modifying the natural order of the world.

The main consideration to take into account for critics of genetic manipulation of species is that we have already spent thousands of years doing these types of experiments and modifying the species we coexist with now. It cannot be argued, of course, that these species are untouchable from a current perspective, because in reality having done so in the past means that we have survived and made it to the present. The success of our farming ancestors and their unconventional view of the natural world, which was up till then inalterable, could be a lesson for those who claim that nothing should be modified.

Chapter 4

Cloning and a Failed Attempt

With certain frequency we hear news that a Russian, Korean or Japanese team is planning, about to or in the process of cloning a mammoth. This news is false, and the scientists involved have no experience working in paleogenomics nor, in reality, in de-extinction. Because those of us who work with remains from the past know that, in the first place and in spite of their appearance, mummified tissues are not favorable for preserving DNA and, secondly, although the remains are found preserved in ice, the genetic material is always degraded into billions of small-sized fragments. Complete cells do not exist, as if they had been cryogenically stored in a biobank that simply allows the nucleus to be transferred and an embryo created to use for cloning.

Some outlandish attempts to insert cellular nuclei from the mammoth into mice oocytes failed immediately, with all certainty for this reason. And it should be remembered that just one mistake or rupture point in the DNA sequence would be lethal. It must be said, very little is known about the reproductive biology of elephants, and an ovule has never been extracted from a female one. Nor is it likely that many operating rooms exist with room for a pachyderm. The fact that the most obvious difficulties are never mentioned suggests that these are publicity stunts that journalists repeatedly fall for. In fact, the relationship between the media and scientists working on de-extinction is one of the biggest problems in this discipline.

The Attempt with the Bucardo

It is true that cloning has emerged as a possible de-extinction system, although only in certain specific situations that we will explore below. In truth, one of the few serious attempts at de-extinction that is always cited happened using the aforementioned technique, although it is a very special case: we are talking about the bucardo, a subspecies of goat from the Pyrenees that we already mentioned in previous chapters and whose last specimen died in

2000. Two peculiarities exist in this case that limit its application to almost any of the other cases listed in the chapter on candidates for de-extinction.

The first is that frozen skin cells (fibroblasts) were preserved from this specimen, which came from a biopsy that was done one year before its death. This does not exist in any other case. As such, this technique becoming generalized is, at the very least, doubtful (a living specimen of the Yangtze dolphin could be discovered, for example, and a tissue sample taken for later cloning, but it is obvious then that the species would still exist, and as such it would not be a case of de-extinction). The other peculiarity is that the bucardo (*Capra pirenaica pirenaica)* is described as a subspecies of the Iberian wild goat *(Capra pyrenaica)*, and here is where we enter into the slippery territory of the definition of this taxonomic category. I think it is increasingly clear that species is an arbitrary concept (although very beloved by taxonomists, paleontologists and conservationists).

A species is, in theory, a reproductively stagnant population, but it is often not like this when two or more of these populations or species meet each other in the wild – or in captivity, where there is a lot of free time – and we discover that they can still hybridize; this is the recent case with the brown bear and the polar bear, or with modern humans and the Neanderthals. Some have tried to argue that they should then be considered the same species, but it is difficult that evolutionary biologists who are experts in phylogeny, conservationists who fight against climate change or the general public swallow this particular pill: call them what you want, a brown bear is different from a polar bear, and the separation of both lineages has been estimated, with genomic data, to be more than one million years.

5. The bucardo and its similarities to the mountain goat

In truth, it is irrelevant what we call these biological entities, as it is really only a question of convenience. Even more doubtful is the concept of subspecies, which is what slightly distinct populations are called within the same species and, as such, can always crossbreed with each other. But in this case, beyond the debate on whether something like a subspecies, which is a category that in reality does not exist, can be considered extinct, this has a huge advantage to attempt cloning because there are surrogate mothers of the same species and, as such, with very similar biology. The fact that, even with these unique conveniences, cloning the bucardo failed only confirms the enormous difficulties that must be faced in virtually every other situation.

Basically, they carried out two series of experiments at the Agrifood Research and Technology Centre of Aragon in 2001, where they transferred the nucleus of a fibroblast from a bucardo to a previously enucleated oocyte from a domestic goat to produce a viable embryo; the nucleus and the oocyte fused by means of electric pulses and created embryos that were clones of the original bucardo (this is a technical procedure inaugurated in 1996 to achieve the first cloned animal, the famous sheep Dolly). These embryos were transferred to female surrogates from the Iberian wild goat subspecies or to hybrids between this and the domestic goat. In the first series, 54 embryos were transferred to 13 females; two pregnancies were produced that were unsuccessful.

In the second series, 154 embryos were transferred to 44 females; they got five pregnancies, of which only one was carried to term. Unfortunately, the bucardo that was born, by caesarean, died immediately due to respiratory problems. Its survival would have been more a symbolic than technological milestone, but nature did not concede even this point.

Although the experiment showed that along general lines the approach was viable, it was not without controversy and was criticized by ecological industries and naturalist entities, to the point that the experiment has not been repeated. An added problem, which did not end up being considered due to the premature death of the specimen, was the fact that it only had cells from a female specimen. To recreate a natural population, crossbreeding it in the future with male Iberian wild goats would have been inevitable.

With the exception of the bucardo, viable cells are not available from any other candidate species. But it is indeed true that the nuclear transfer technique should be used in the case of attempts with genetic editing or synthetic biology, at the end of the process. That is, you could take cells from a living relative of the species, modify them with genetic engineering in the laboratory to add specific genetic variants from the extinct species, and then transfer the nuclei of these cells to enucleated oocytes to create embryos that would need to be implanted in a surrogate mother. It seems to be the easiest part of the process, but it is not. As we will see, it could be that our problems have only just begun.

The Problem with Surrogate Mothers

In de-extinctions, we must always find a different species that would act as a surrogate mother. A few million years of evolutionary divergence implies insurmountable reproductive barriers. This difficulty explains the failed attempts by Russian doctor Ilya Ivanovich Ivanov in the 1920s. Ivanov carried out a series of controversial experiments to create a hybrid human-chimpanzee. First, he tried to inseminate female chimpanzees with human sperm, but he was unable to produce a pregnancy. After that, he searched for a volunteer to be inseminated with orangutan sperm, but this experiment never took place due to the animal's death. Ivanov was a victim of the purge in 1930 and sent to Alma Ata, where he died two years later. Ivanov's motivations are controversial; although it has been suggested he acted following Stalin's orders, and that his objective was to create a race of "super-soldiers," it appears that in reality he wanted to show by means of hybridization that human beings and present-day pongos shared a recent common ancestor. In any case, we can ask ourselves what are the limits in order for genetic and physiological incompatibilities to make a possible surrogate mother unviable.

Various examples exist of interspecies somatic cell nuclear transfer (iSCNT) that have had positive results, but always between very close species. The coyote *(Canis latrans)* has been cloned using the dog *(Canis lupus familiaris)* as a surrogate mother, the African wild cat *(Felis silvestris lybica)* and the sand cat *(Felis margarita)* using the domestic cat *(Felis catus)*, and the mouflon *(Ovis orientalis musimon)* using the domestic sheep *(Ovis aries)*.

In some cases, the procedure has been used to try to save a threatened species, such as the gaur *(Bos gaurus)*, a large bovine native to southeast Asia and India. The attempt was made in 2001 with a domestic cow *(Bos taurus)* as a surrogate mother, and ended with the death of the cloned specimen 48 hours after birth. The death was attributed to a bacterial infection unrelated to the cloning itself, and repetition of the experiment produced, some years later, a healthy guar. Some, however, have criticized this specific project; with a population estimated at 36,000 specimens, it could be argued that the gaur's survival has more to do with preserving its habitat than with cloning.

It is hoped to use a similar approach to save the subspecies of northern white rhinos that were formerly found throughout central and eastern Africa, and of which now only three live specimens (two females and a male), who apparently are sterile, remain. The two surviving rhinoceros, named Najin and Fatu are mother and daughter (the father and grandfather and last male, Sudan, died in March 19, 2018), belong to a zoo in the Czech Republic but, in fact, live on the Ol Pejeta Conservancy in Kenya, where they are watched day and night by armed guards (poaching, focused on their horns, is what has decimated rhinoceros populations in the last decades). One of the possibilities being considered, following numerous failures with assisted reproduction techniques, is to implant an embryo by means of nuclear transfer in a female southern white rhino (whose population is around 20,000 specimens), an attempt that will be carried out at the San Diego Zoo.

These experiments show that the technique is possible, but it is also evident that some problems exist having to do with the viability of the descendants. In part this is due to the lack of knowledge we have on the reproductive biology of many current species. One unfavorable aspect is that the efficiency of viable embryos with the aforementioned technique, due to incompatibilities between the nucleus and the cell, is always very low; it is estimated that between one and six percent of the embryos created can come to term. This raises the price of the process, as you need to manage a large number of experiments and pregnancies.

Furthermore, it would be necessary to investigate, in the hypothetical case of a successful birth of a healthy animal, if the cloning process might have caused differences in physiology or behavior when compared to these aspects in members from before the extinction. One of the problems is something known as phenotypic plasticity. It has been seen that small and subtle differences that occur during development, such as thermal or chemical factors, can notably influence external aspect of the members. The fact of having to use the cellular machinery from another species for cloning could result in the appearance of abnormal phenotypes, which has been observed, for example, in fish.

In cases where there does not exist sufficiently close species that can act as surrogate mothers, a potential solution for the future is the development of artificial uteri. It must be said that this area of research is still in its infancy, so it is difficult to predict how and when *in vitro* development of offspring will be possible. Although tests have been made with mouse embryos that have survived a week (half of their natural gestation period), the truth is that, as Helen Pilcher, a scientific journalist from the BBC, said, "the technology for artificial uteri is still, well, embryonic." A possibility that could resolve some problems would be the use of real placentas from available animals in artificial uteri.

Cloning problems do not end here; it is worth remembering that every list of de-extinction candidates has birds and occasionally reptiles and amphibians, in addition to the obvious mammals. It turns out that because of particularities in the reproductive physiology of birds, whose fertilized oocytes cover themselves in a resistant shell before being laid, genetic engineering techniques are much more complicated than with mammals. In fact, it has scarcely been tried. When the egg has the shell formed, the embryo is in too advanced a phase of development. One suggestion has been to extract the oocytes and cultivate them in albumin, but attempts made with chickens have confirmed that it is impossible to see the nucleus and, as such, manipulate it. Genetic editing techniques seem just as complicated, given that preservation of a cell as big as an egg for the necessary years needed to be modifying a significant list of genomic positions could be impossible.

These barriers are a real challenge for some of the most obvious candidates for de-extinction, such as the great auk or the passenger pigeon. As a possible alternative it has been suggested resorting to manipulated blastoderm cells that would be added to the ovaries; this would result in chimerical animals that could produce eggs derived from the previously modified external cells. If two individuals were produced with this method and were crossbred, it could result in an animal that now only had the modified DNA. All the cloning problems we have seen in the case of birds also applies to reptiles, given their physiological similarities.

Conversely, with fish there have been positive results from experiments in nuclear transfer and cloning, so they could be easier candidates for de-extinction (although they are animals that are traditionally forgotten on all the lists). Working with amphibians seems equally as promising; in fact, the first cloning experiments were done on frogs in 1952, and also the first cloning via transfer from a somatic cell in 1958. A possible candidate for de-extinction would be the Australian gastric-brooding frog *(Rheobatrachus silus* and *Rheobatrachus vitellinus)*, extinct in 1980, which was characterized by ingesting its eggs and incubating them in its stomach until live baby frogs were born from its mouth. Epidermal cells are preserved from some specimens and, in an experiment in 2013, Australian scientists were able to create embryos by means of nuclear transfer from these cells, although they went no further than the tadpole stage.

In conclusion, due to the limitations associated with cloning as the only technical strategy, alternative methods for de-extinction are starting to be explored, which would involve genetic editing or the creation of synthetic genomes. Both approaches have difficulties that are not trivial, but some people believe these could be resolved in the future.

Chapter 5
Synthetic Genomes and Genetic Editing

It is understood as synthetic those genomes that are functional but have been created *de novo* in the laboratory by means of a process uniting short chains of oligonucleotides until they form a more or less long sequence of DNA. And synthetic biology is defined as a discipline based on the chemical synthesis of DNA to respond to human needs by creating organisms with new or better characteristics. It is a technique that is still in its infancy and, at the moment, is facing considerable challenges to be able to go from the limited scale of a bacteria to what would approximate a standard genome from a vertebrate. Nevertheless, thousands of millions of euros are invested every year in synthetic biology, and many perceive the field as transformative technology that will fight against climate change, the scarcity of energy, water and food, as well as future challenges having to do with security. What emerges from the interaction between synthetic biology and society is in reality a "bio-economy."

Ever since the Nobel prize-winner H.G. Khorana synthesized a gene with 207 nucleotides in the laboratory in 1979, the size of synthetic genomes has increased exponentially to reach sizes of around one million nucleotides. But the road has not been easy.

The Road to Synthetic Organisms

In an initial study carried out in 2008 at the J. Craig Venter Institute (Venter being the controversial father behind the private Human Genome Project), they synthesized the 583,000 nucleotides from the genome of *Mycoplasma genitalium*. To do so, they used *in vitro* enzymatic recombination methods and *in vivo* homologous recombination; these laboratory techniques use enzymes to cut DNA chains and join them with other different ones. The first method was used to unite 25 DNA sections with lengths of between 24,000 and 72,000 nucleotides, which later were joined in four sections of 144,000 nucleotides each. The efficiency of *in vitro* recombination declined as the fragments became longer to the point that it was no longer possible to continue assembling fragments, so it needed an *in vivo* recombination method using yeast *(Saccharomyces cerevisiae)* for the final assembly. Basically, the study was a description of the methodology required for the synthesis of artificial genomes.

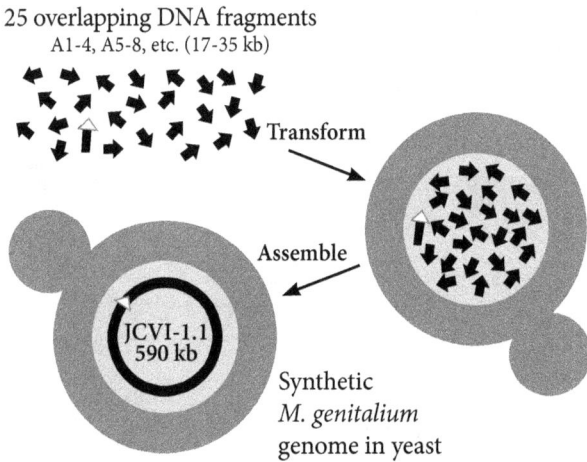

6. Diagram showing how to create a synthetic genome from mycoplasma using yeast to assemble the fragments

In 2010, the same researchers manufactured a synthetic genome of a bit more than half a million nucleotides from *Mycoplasma mycoides* that they inserted in a *Mycoplasma capricolum* from which they had previously eliminated the genetic material; the resulting cell showed the functions that had been previously designed artificially and was able to live correctly by managing the cellular machinery.

The researchers, which included the Nobel-winner Hamilton Smith and Venter, called the resulting organism *Mycoplasma laboratorium*, to emphasize its *in silico* nature. The initial tests failed, and for various months they had to look for the source of the problem; finally, they discovered that just one mistake in the DNA sequence had caused a protein intervening in genetic translation in the ribosomes to be defective. This illustrates the technical difficulties of these experiments, and the precision required for them to work.

To emphasize their nature of being designed by human beings, the researchers inserted into the DNA sequence four encrypted messages with phrases from celebrated people and also a script in HTML code that, once put into a browser, created a greeting for the decoder and a link to click on showing that they had indeed decoded the code. Among the messages was an evocative phrase from James Joyce: "To live, to err, to fall, to triumph, to recreate life out of life."

The study on synthetic *Mycoplasma*, which was presented to the press as an artificial creation from a synthetic organism, was an object of controversy in the scientific world for the fact that, in truth, it needed a live surrogate organism to be able to function, something that was ignored in the media coverage of the news. For Jay Keasling, a pioneer in synthetic biology, "the only regulation we need is of my colleague's mouth" (referring to Venter).

An additional debate concerns the fact that the

J. Craig Venter Institute patented the genome from *Mycoplasma laboratorium*; patents on biological organisms – and it would be necessary to debate whether this is the case here – are controversial, and there are diverse entities opposed to them. It is estimated that the cost of the entire project was around 40 million dollars. But thanks in part to the media impact, Venter was able to collect almost three times as much money for a project to create synthetic algae to make biofuels, in collaboration with Exxon Mobil.

After this, the same researchers decided to take a step towards eukaryote organisms (that is, organisms that have genetic material from a cell nucleus enclosed within a membrane). First, they were able to synthesize the complete mitochondrial genome (as it remains in origin a bacterial genome) from a mouse, uniting 600 fragments from 60 nucleotides until reaching a total of 16,300. In 2014, a complete eukaryote chromosome of around 270,000 nucleotides, also from yeast, was synthesized for the first time; the *in vivo* recombination technique was used, done by replacing increasingly larger sections of the original chromosome with DNA chains synthesized in the laboratory. On March 10, 2017, work published in the magazine *Science* showed the synthesis of five more chromosomes from yeast (which raised the key genetic material from this artificially synthesized organism to 30%). The consortium from the Synthetic Yeast Genome Project announced its plans to finalize the synthesis of the 16 chromosomes that make up the yeast genome.

It should be remembered that synthesis of eukaryote genomes is much more difficult, not only because of their greater length but also for the complexities of the internal compartmentalization of the eukaryote cell. Although what has been achieved to now represents less than 0.01% of the total length of the majority of animal genomes, and continuing down this road with current techniques would represent considerable costs in both time and money, it is possible that the techniques will improve in the future. In particular, it is expected that at least the most tedious parts of the laboratory work can be automated; needing much less technical personnel means that part of the process can be done more economically.

But synthetic biology does not only aspire to creating *de novo* genomes. George Church, a renowned expert in genetics and synthetic biology from Harvard Medical School in Boston, is even trying to transform the universal genetic code, which as we know is restricted to only 20 amino acids. The incorporation of new amino acids could allow the creation of new proteins with unknown functions. Some of the ideas being considered include the design of bacteria that can degrade toxic substances in the environment, eliminate tumors or synthesize biofuels and new materials. There are projects that seem promising, such as the modification of *Escherichia coli* (a bacteria that lives in the intestinal tract) so it secretes a peptide that is involved in the feeling of satiety. As such, it would be possible to override the appetite in obese people without needing to do aggressive treatments such as stomach reduction or to go on a diet.

Future Challenges in Synthetic Biology

The applications of synthetic biology on industry in the future are evident. In part for this reason, synthetic biologists talk like engineers – and in fact, at times they are. Instead of speaking about genes and metabolic routes, they prefer terms such a parts, devices and modules. A "part" is the coded region of a gene and its regulatory system for genetic expression; a "device" is a set of parts that together carry out a certain function, such as to "turn on" or "turn off" a gene, and a "module" is a set of devices that direct more complex tasks such as coordinating a certain chemical route. The physical nature of DNA can even be likened to the transistors that make up microprocessors, and the DNA sequence to computer programming code. But all these engineering abstractions do not change the fact that, in reality, all the most elemental technical questions need to be resolved in a damp laboratory, which is where experimental biologists work.

Even so, synthetic biology represents one of the greatest conceptual leaps in current knowledge, and there is no doubt that it will make a mark on society in the future. Craig Venter's response to critics who said that the synthesis of artificial genomes would not have any practical application was: "It's like asking why do you want to build an airplane when we already have horses?"

Synthetic Biology and De-extinction

Nevertheless, synthetic biology applied to de-extinction must negotiate some additional challenges in the future. For example, the reconstructed genome must be sufficiently complete to be functional, a complicated thing when dealing with genomes from species that have disappeared. This difficulty is without a doubt a challenge, because due to fragmentation of the original DNA, when we sequence a paleogenome we must always do it by mapping the DNA sequences obtained on a reference genome. The closer the species used as a reference, the more exact the mapping will be. For example, I am currently collaborating with George PJ Perry, a biologist from Penn State University, to recover the genome from an extinct lemur, *Megaladapis edwardsi*. The phylogenies done with mitochondrial DNA indicate that its lineage separated from other Madagascar lemurs around 30 million years ago; this implies not having a good reference available to map the recovered sequences, since evolutionary changes have accumulated in one or the other lineage during a not insignificant period of some 60 million years. In practice, it is much more complicated to map two species of lemurs from the same African island than it is to map the human genome with that of the mouse. It is almost certain that we have sequenced the complete genome of *Megaladapis*, it is simply that we cannot know.

Be that as it may, the mapped genome will never be exactly identical to the original, and will never be all of the genome, given that the numerous repeated sequences, which can represent 30% or more, cannot be mapped under any circumstances. This is not especially relevant for the genome's functionality, although synthetic chromosomes indeed must have centromeres that are similarly designed to existing ones, as these intervene in cell division. More worrisome is the fact that it would be very difficult for us to recognize chromosomal reordering or regions that have evolved more rapidly and which, as such, would show very divergent sequences between both species. This would be serious precisely because they would belong to genes that are important in defining the extinct species. What would happen is that these "new" sequences would not map in the reference genome, and we would discard them as sequences that are environmental or contaminating. The consequence is that the de-extinct species will never be totally identical to the original. But on the other hand, it is true that this also happens in other methodological approaches (for example, in the sequencing of current organisms), and it is difficult to control.

Beyond the linear DNA Sequence: the genome architecture

In addition, attaining ancient genomes to synthesize must take into account questions having to do with genomic architecture, that is, it will be necessary to design beyond the DNA sequence of protein-coding genes so that the genome created will be functional within a cell. This means that the regulatory sequences distributed by the thousands throughout the genome, often physically far from the gene whose expression they control, must also be known.

Nevertheless, one of the current fields of study is the design of a standard set of regulatory sequences that, in an established way, control the expression of any protein. In a future genome designed synthetically, it would then not be necessary to reproduce the regulatory sequences from nature, but rather they could use those from the catalog, according to the needs of the designers.

On the other hand, some researchers believe that a large part of the intron sequences – non-coding sequences that separate the coded parts from the genes – could be eliminated in a synthetic genome, as well as redundant or non-coded parts (what had traditionally been called "junk DNA") and which make up the immense majority of the genome. That is, a genome of various gigabases (thousands of millions) of nucleotides could work the same as one with a few dozen million nucleotides. Without a doubt, this would lower costs and smooth out the difficulties. An existing example is a bacteria strain *Escherichia coli*, called MDS42, from which they have eliminated all the unessential elements, such as pseudogenes (unused genes that are preserved like genomic fossils) or virus sequences, which has resulted in the elimination of 15% of the genome. In the synthesized yeast genome, nearly 8% of its length has been eliminated to date. A eukaryote genome could be reduced in an even higher percentage.

Additional problems related to genomic architecture refer to a lack of knowledge on how the DNA chain is packaged, forming complex proteins within the cell, and which mechanisms control this process. It should be remembered that these structures affect cell transcription and replication, so an erroneous or incomplete ordering would without a doubt have consequences in development of the organism. It has been seen how exogenous DNA inserted into a frog cell spontaneously forms structures similar to nucleosomes, but it is not clear how it would be possible to functionally structure multiple synthetic chromosomes. Also, the same genome functions differently in different tissues, and a large part of the mechanisms in cell differentiation are not well understood.

In conclusion, perhaps the biggest problem will not be to design and synthesize the genome, but rather to put it to work. Undoubtedly, the complexity of the biology is much greater than the chemical synthesis involved in making a DNA chain. But when synthetic organisms are available, we must imagine that they will interact with live species and, in turn, evolve. Where there is change and reproduction, over time there is evolution, and it remains to be seen how these new organisms will affect natural ecosystems. Incidentally, the terms "natural" and "artificial" will soon make no sense, and we will also have to see how public perception of these concepts changes, as well as the idea of evolution as a process not under human control.

Up to now, conservation biology has ignored synthetic biology; but independent of possible achievements from the latter in cases of de-extinction, it is evident that its development might affect the first discipline. For example, synthetic biology could find solutions to provide food to a future world population of more than nine billion people, which would indirectly mean alleviating pressure on natural ecosystems that are currently being destroyed to be turned into arable land. Conservationists have relied on quite traditional practices to ensure that diversity in the wild is maintained, and the truth is that their success with this task has been rather limited. Perhaps it is time to accept new technologies as being more imaginative tools, without falling into triumphalism or alarmism. The hope exists that collaboration between the synthetic biology community and that of biodiversity conservation – up to now strangers, both in their own fields – may change the relationship between the natural world and humans.

Genome Editing

An alternative to reinventing extinct species, potentially more viable than synthetic genomes, is to edit the genome of a related living species using the genome sequence from the extinct species as a guide. The nearest example we have available is genome editing of the Asian elephant with mammoth genes, something already being done in George Church's laboratory, but the principles used here are applicable to other cases. Even so, it is possible that we have begun with the most difficult case, and working with smaller animals would be much easier and more viable.

The CRISPR Discovery

Genome editing is only possible thanks to recent developments that have taken place with the technology known as CRISPR-Cas9 (this is an acronym for *"clustered regularly interspaced short palindromic repeats,"* a term proposed by the researcher from Alicante Francis Mojica, who studied them in 1993). The CRISPR system is an immunity mechanism present in prokaryotes and archaea that provides resistance to external agents such as plasmids and virus. The CRISPR compound comprises sets of repeated sequences of 24 to 48 nucleotides, distributed along the bacterial genome and associated with Cas genes, which are codified for nucleases.

Basically, when a virus penetrates into a bacteria of the CRISPR, the union of the intruder's genetic material with the Cas enzymes and some RNA sequences from the system provoke inactivation and degradation of the virus. But additionally, part of the virus sequences are integrated into the CRISPR system in such a way that if this bacteria again enters into contact with the same virus, it will inactivate the virus much more rapidly and efficiently. Years after its discovery, some biologists realized that this system, because it cuts the DNA in specific positions of the genome, could be used for precise genome editing, something that was impossible up to then.

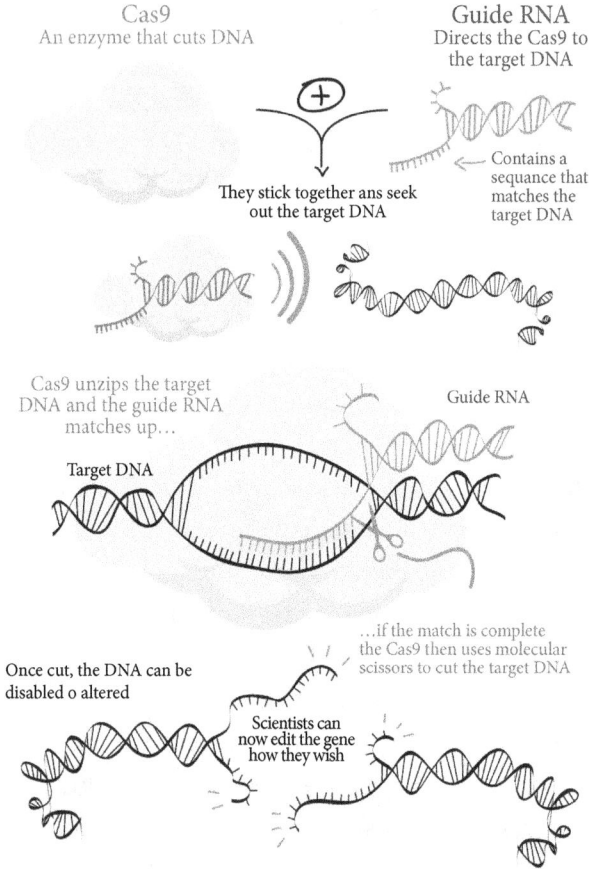

Cas9
An enzyme that cuts DNA

Guide RNA
Directs the Cas9 to
the target DNA

They stick together ans seek
out the target DNA

Contains a
sequence that
matches the
target DNA

Cas9 unzips the target
DNA and the guide RNA
matches up...

Guide RNA

Target DNA

...if the match is complete
the Cas9 then uses molecular
scissors to cut the target DNA

Once cut, the DNA can be
disabled o altered

Scientists can
now edit the gene
how they wish

7. Diagram of how CRISPR functions: modifying genes

In its application to genome editing, two key
pieces of the CRISPR system are used: the Cas9
enzyme, which cuts the double DNA chain in specific
points, and an RNA, called simple guide RNA (or
sgRNA), which is a short sequence designed to
link up in the specific point to be edited (because
it is complementary to the DNA sequence from the

individual) and which guides the Cas9 enzyme there. Without the guide RNA, the enzyme could cut the genome in unwanted points. Once the cell detects that there is damage to the DNA chain, it tries to repair it, which then allows a specific DNA sequence to be introduced in that precise position and with the mutation you wish to incorporate into the organism.

Editing of Current Genomes

The high specificity of the system has led it to be used in questions of genome editing, to correct "mistakes" in genes that produce hereditary diseases. For example, the most common mutation causing cystic fibrosis (the most frequent and harmful autosomal recessive disease in Europe) is deltaF508, and it is a deletion of only three nucleotides in the *CFTR* gene. On paper it is possible to edit the gene with the correct version of *CFTR* and reintroduce it in the bone marrow of the patient, who will begin to produce healthy cells. It is estimated that replacing around 70% of defective cells would represent in practice the recovery of the patient, currently a person with a life expectancy of from 37 to 40 years.

Nevertheless, medical use of CRISPR-Cas9 is still in its first clinical trials, as some problems have been detected; in some cases, the cuts in the DNA chain are done in undesirable areas of the genome (with the fatal consequences this can entail). It has been seen in diverse experiments that the range of successful cuts is low and quite variable, going from 0.5 to 20% in the best of cases. The inefficiency observed and the fact that each cut requires subsequent controls to check that it has been made in the correct spot show that the system is still very inefficient if what you are after is to edit a large number of positions in just one genome. But in some clinical trials it has been shown that the alternative to possible dysfunctions from CRISPR is the inevitable death of the patient, and promises from genome editing, their only hope.

Genome Editing and De-extinction

Here we reach a key question, which is: how many genome positions must be edited for a potential de-extinction? We have already seen that between mammoths and Asian elephants there are, at least, changes in more than 1,600 proteins; but in addition, it will possibly also be necessary to alter hundreds of regulatory sequences, beyond the amino acid changes in the genes that code for proteins. A genome has various levels of complexity, and proteins are only the most elemental and obvious of evolutionary differences. Because an organism is not only its DNA sequence, but also the result of the control that regulatory mechanisms exert on the final expression of gene products. There are mutations in regulatory sequences, for example, that make it so that a certain gene transcribes and translates into protein with greater intensity. It would also be necessary to silence 26 genes that are active in the Asian elephant but inactive in the mammoth; this must be done by altering the regulatory region or inserting a STOP codon in the middle of the gene, but in any case it would require additional editing to active changes from the mammoth lineage. Because being a mammoth also consists in not being an Asian elephant.

Even when milestones in genome editing appear to be of reasonable magnitude, carrying it out successfully is still an enormous enterprise with our current state of knowledge. It must be remembered that, to date, the majority of studies based on CRISPR have focused on just one or very few insertions in the genome, and this alone has been an enterprise of considerable scope.

Building of the Mammophant

We must speak about George Church again here, a controversial character (you only need to remember his declarations about cloning a Neanderthal using a modern human as a surrogate mother) and for some, without a doubt, brilliant, who is among the vanguard in genome editing with the goal of recreating a mammoth in the laboratory. Church claims that thousands of changes are not necessary to obtain an animal that can meet the ecological functions of a mammoth, and that it would probably be enough with a few dozen key changes. It is true that it is very probable that many of the changes detected on the DNA sequence do not have a perceptible effect on the resulting proboscidea, but it is also true that to make a list of the significant changes may imply as much work – by means of functional studies – as editing them and seeing what happens.

A separate question will be to debate whether an Asian elephant with a few dozen mammoth mutations, no matter how important, represents a true de-extinction. In truth, this approach is more similar to a fast way of carrying out in one fell swoop the different and tedious functional studies; in a certain way, it is to de-extinct some genes instead of a species.

Although this work has still not been published scientifically, it seems that Church already has Asian elephant cells with 45 genetic changes from the mammoth, in principle obvious cases of adaptation to the cold, like the already-known hemoglobin or genes related to lipids and fur. To be able to verify the real effect of these changes, Church proposes changing the fibroblasts into pluripotent stem cells – which are immortal in the laboratory – and trying to develop them in different organelles. An obvious step is to obtain blood cells with these that can be used to measure affinity with oxygen at different temperatures. Another, like differentiation in follicular cells, may be more complicated to achieve.

Once these mutations have been tested *in vitro*, Church believes that it will be possible to reach the step of transferring the edited nucleus to an elephant oocyte and to wait a couple of years (the time gestation lasts for an elephant) to see what happens. His projections situate the first attempts to create a mammoth embryo (or a "mammophant" embryo to be more precise) in 2019, but these do not appear to be realistic. In fact, the announcement of this calendar, made by Church at a conference of the *American Association for the Advancement of Science* in February 2017 in Boston, was directly labeled as "false news" by the paleontologist and blogger John Hawks.

The Editing of Paleogenes in Laboratory Mice

Some of the changes studied could be simplified thanks to the use of laboratory animals. It might not be necessary to create stem cells if sequences from extinct animals could be inserted in mice to see their effect. In 2008, a study was published that in a certain way proposed a vision, although limited, of this approach. The researchers recovered a regulatory sequence from the *Col2a1* collagen gene from four Tasmanian tiger remains, coming from three offspring conserved in alcohol and from fur of an adult specimen from the Victoria Museum in Melbourne. They wanted to prove, beyond the mutations in genes that code proteins, how some genome elements could influence the transcription differential of genes in extinct species (just as they do in live species).

First, they confirmed that the sequence from the regulatory element selected was similar, but not identical to, that of other marsupials. Next, they joined it to a transcription reporter, the *lacZ* bacterial gene. In molecular biology, reporters are sequences that attach to the gene it is hoped to study, as the characteristics they give to the organism where they are inserted are easy to detect and quantify. In the case of *lacZ*, it turns an intense blue color when it grows in a medium with an organic compound known as X-gal. In this way, it is possible to know where the *lacZ* gene is being expressed (which, we must remember, we have attached in this case to the *Col2a1*). The researchers from the thylacine study created diverse mice embryos and after interrupting the gestation after 15 days, they dyed the resulting fetuses with X-gal and were able to observe that the regulatory sequence of the thylacine, as was to be expected, was being expressed in diverse areas of the skeleton, where there were developing chondrocytes. The pattern was identical to that observed in the regulatory sequence from the *Col2a1* gene from a mouse, which indicates that the sequence from the thylacine performed the same function in spite of both lineages having diverged from each other some 148 million years ago.

Although it cannot be qualified as spectacular, this study represents a step beyond previous functional studies done *in vitro* (such as those carried out with the MC1R protein or the hemoglobin from mammoths), since it provides evidence within a living organism. Nevertheless, any gene from an extinct species that you want to study using laboratory mice can have its interpretative complications, owing to the evolutionary distance involved. That is, it could be that the same regulatory sequence – or one same gene – in a thylacine ends up expressing itself in a different way in a mouse. Researchers that work on human diseases and use mice as model animals face the same problems. But evidently it is necessary to live with these limitations in both cases, given that we cannot experiment directly on humans, nor can we do so with species that have disappeared.

As we can see, the challenges that all these technologies combined are facing are still enormous but, on the other hand, recent and unforeseen developments such as CRISPR have relaunched the field of de-extinction, to the point that the leading scientist at "Revive & Restore," Ben Novak, has said that his project to revive the passenger pigeon "is 100% impossible without CRISPR." Be that as it may, the CRISPR zoological has begun its journey in science and in society, and everything points to it being a fundamental technique, not only in de-extinction but also for the future of medicine.

Chapter 6

Selective Breeding and the Quagga

Of the four methods proposed for de-extinction, there is one, selective breeding, that does not require any technological advance; in truth, it does not require any advanced scientific knowledge because humanity has spent thousands of years doing it, first with domestication in the Neolithic and, more recently, with the creation of different breeds of dogs, sheep or pigs. As we have seen, the characteristics that have traditionally been selected have been related to less aggressive behavior, resistance to diseases, larger size, better meat, greater milk or wool production, higher fertility, etc. Darwin was aware that this type of process was similar to natural selection, and he developed this concept in two chapters of the original edition of *On the Origin of Species* (in fact, he coined the phrase "selective breeding"), as well as in a specific book titled *The Variation of Animals and Plants under Domestication*, published in 1868.

This idea deals with selecting, generation after generation, those individuals that have the most pronounced, totally or partially, external feature we want to fix in the population, and which will increase in frequency in descendants. In essence, it is about artificially controlling the reproductive biology of these animals and deciding which will have descendants and with whom. As we have previously seen, this strategy can only be used with animals that can adapt to conditions in captivity.

The more unusual the trait we seek, the greater the probability that the new variety comes from few original individuals, with the added problem of starting from low genetic diversity and with problems of endogamy. For example, the American and European stock from the sphynx breed of hairless cat, created beginning in 1960, comes from only two litters and five original individuals, two from Minnesota and three from Toronto.

Obviously, selective breeding is not the most spectacular method, and is occasionally being used with the goal of de-extinction. In these cases, an external physical trait is looked for that is reminiscent of the extinct organism. This has the inconvenience of coming from the outside, from the exterior of the animal and ignoring the rest of the information. We could say it is an essentially visual process that in no way ensures that the genetic basis of the extinct species has been brought back to life. At best, we can expect that mutations in the gene or genes involved in the trait or traits chosen have indeed again become predominant in the resulting population. The most well-known project is the one with the quagga, although others are underway, such as with the auroch *(Bos primigenius)*, the ancestor of the wild cow.

8. Sphynx cats are characterized by their lack of hair

The Rebirth of the Quagga

The quagga *(Equus quagga quagga)* is found in the very beginnings of paleogenetics, given that the recovery of some fragments of its mitochondrial DNA in 1984, coming from a naturalized specimen from the Natural History Museum in Mainz, is considered the official beginnings of the discipline. It was a species or subspecies of the plains zebra that lived south of the Vaal River in South Africa. It was different from other zebras because it had stripes only on the upper part of its body. The quagga was hunted excessively by Afrikaner colonists during the second half of the nineteenth century, to the point that the last known specimen died on August 12, 1883, in the Amsterdam Zoo.

Preliminary genetic results, confirmed in 2015 when the complete quagga genome was attained, indicated that it was very close, phylogenetically, to what was known as the Grant's or plains zebra *(Equus quagga boehmi)*, to the extent that it could be considered a subspecies or subpopulation of this zebra. In the genome work, directed by Ludovic Orlando at the Centre for Geo Genetics in Denmark, it was estimated that both equine lineages had diverged between only 233,000 and 356,000 years ago.

Years before knowing even the first genetic results, the German zoologist Lutz Heck (creator, along with his brother Heinz, of the Heck cow) suggested, in 1955, that a careful selection of selective breeding between plains zebras could produce something very similar to the extinct quagga. This idea was picked up again years later by the German naturalist Reinhold Rau (1932-2006) who founded the "Quagga Project," which was based on the same concept. The promoters of the project examined 23 quagga furs preserved in different museums and observed that there was a certain variation in the number and dispersion of the stripes and, as such, believed that the genes of the quaggas still survived, although diluted and dispersed, in current zebra populations. As you can read on their website, the goal of the project is to use selective breeding to try to obtain a population that in their external appearance, and possibly also in their genetics, is close, if not identical, to the ancient population known as "quagga."

The first real steps of the project took place in March 1987 with the capture of nine specimens of plains zebras selected in the Etosha National Park in Namibia, which were later transferred to a fenced area on a farm near Robertson in South Africa. They began crossbreeding between the members with less stripes, discarding at times those that did not present a clear pattern (these animals were set free in Eddo Elephant National Park) and incorporating new specimens they captured in the wild. As the number of zebras increased, the farm became too small and the project had to be transferred to a larger area. On June 29, 2000, the director of the Quagga Project and the director of South Africa's national parks signed a cooperation agreement, and as such the initiative passed from private hands into public ones.

The Return of the Quagga

On January 20, 2005, the first zebra was born, named Henry, who could be considered externally indistinguishable from original quaggas; after only a few years, the fifth generation of selective breeding and 116 individuals within the project was reached. The members who are born with the pattern of reduced stripes – as of now 11 are listed, named Henry, Freddy, DJ14, FD15, Nina-J, Khumba, Julian, JT16, RM16, TS16 and François – are called "rau-quaggas," in honor of the promoter of the project. In the near future they hope to reach around 50 of these specimens, at which point they will be set free in a controlled area of their natural habitat.

The webpage of the Quagga Project seems to be unaware that the quagga's complete genome is already available, and as such it would be possible to analyze how many of the genetic variants of the original quagga are present in the rau-quaggas. It is logical to think that we can only hope to find some variants related to the shape of the stripe pattern, but not others that could be involved in adaptive aspects that are metabolic, behavioral, physiological or dietary. In this sense, the authors of the genome have detected that there is a gene, *SEMA5A*, that has been specifically selected in the quagga lineage but not in other zebras. It is not known exactly what this gene does; it is a nerve axon guidance factor that in studies on mice has been described associated with the cranial vascular pattern and the volume of the hippocampus. In humans, mutations in the same gene have been described associated with autism. To know their possible function in an extinct equine, it would be necessary to do laborious functional analyses like those done for *MC1R* or *TRPV3* from mammoths. But the genome study suggests that it was an important trait – perhaps behavioral – in the design of a quagga, and the "rau-quagga," who derive from the plains zebra, most certainly do not have it.

The Return of the Auroch

Another example of selective breeding are diverse projects in Europe that are trying to recreate the original aurochs; in this case, it is what we could call a case of de-domestication.

Aurochs differed in diverse traits from their domesticated descendants in the Middle East who entered the continent during the Neolithic: they were of larger size, had huge horns that could reach over 30 inches in length and had hooves that were larger and slimmer than current cows. In the last century there have been various initiatives of selective breeding, some of which have produced varieties, such as the Heck cow in Germany, more similar to the auroch. In recent decades, different projects, such as the Tauros

Programme, have begun the task of selecting and crossbreeding "primitive" cow varieties to produce specimens more similar to the auroch, and in Poland a paleogenome and synthetic biology project has been started, supported by the government, to return something similar to an auroch to the country's forests.

In 2015, a team of researchers sequenced the complete genome of a British auroch from 6,750 years ago, coming from the Carsington Pasture Cave in Derbyshire, and found that it shared part of its ancestry with current Irish and British herds, especially with those from Wales and Scotland. This could be attributed to the fact that the domesticated cow that arrived to the British Isles was perhaps bred with local aurochs to improve their adaptation to a very different climate from that in the Middle East. The results also suggest that perhaps there was a large genetic diversity among European aurochs, and maybe it should be asked which auroch specifically it is hoped to make de-extinct. But the most interesting part of the work is that it created a list of genes that differed between the auroch and the cow, and which had been selected in the domesticated version. These include genes involved in neurobiology, growth, metabolism and immunity. Evidently, the ancestral variants for these traits are no longer found in current cows, and new aurochs will not look like the ancient ones because of these genes, which are key. There will only be an aesthetic similarity in, for example, the length of the horns. But an auroch was quite a bit more than just a pair of long horns.

Another similar project is that of the tarpan, which was a Eurasian wild horse *(Equus ferus ferus)* and whose last specimen – which was already probably a hybrid – died in captivity in Russia in 1909. In the 1930s, various attempts were carried out to obtain a tarpan by means of selective breeding, which sought to recreate some of its characteristics, such as a grey color (with a stripe down its back and darker hooves), small size and strong build. Once again, the Heck brothers took on the job and created what is known as the Heck horse. They made similar efforts with konik horses in Poland, some of which were later introduced into the Bialowieza forest, the last bastion of primeval forest in Europe.

Those responsible for selective breeding seem to ignore the problems of genetic ancestry, in part because their teams are led by experts in reproductive biology and rarely include paleogeneticists. Greater collaboration between both collectives could improve the results of these projects. For example, phenotype selection could be combined with genetics. Genotyping of all the current animals selected could discover that some of them, independently of their external look, were carriers of an ancestral genetic variant in some less-obvious but equally important characteristic. Those individuals that share greater chromosomal fragments with the auroch or with primitive horses, more than specific traits, could be selected; eventually, these specimens would start appearing as the genome base from ancestral species keeps increasing in the descendants. Efforts could be combined with limited genome editing in the case of variants that were not present in the current genetic patrimony. It would then be a much more objective process than attempts guided only by morphology.

With the current focus, the great handicap in selective breeding is the following: the resulting individuals may have the "correct" look, but beyond a few genes, it is evident that we are not selecting anything more than the genetic architecture from the ancestral animal.

Chapter 7

The Ecological Question and the Ethical Question

Everyone is aware that one of the added difficulties in restoring an extinct species is the possibility that its ecosystem, its natural environment and ultimately its food resources no longer exist today. This problem is evident in the case of mammoths and other animals from the Eurasian megafauna, such as the woolly rhinoceros and bison, as the area where they lived is now one of the most unproductive on the planet. But as we will see, this could be a surmountable difficulty.

On the other hand, experts in de-extinction argue that the main objective of such research is not to have extinct species back in zoos, but rather to restore and rebalance ecosystems. Some species play fundamental roles in a certain ecosystem, while others may be redundant (that is, from an ecological perspective other species could do the same tasks). Experts point to two cases in which the recently disappeared species was unique from a functional point of view, and once they became extinct their environments changed dramatically: the passenger pigeon and the mammoth.

The Ecological Role of the Passenger Pigeon in North America

It is estimated that at the beginning of the nineteenth century there were enormous flocks of passenger pigeons, and that their population reached nearly five billion specimens. Their excrement was flammable and so massive that it provoked natural fires that contributed to periodically revitalizing forest areas and increasing diversity. Additionally, they formed the main way of dispersing seeds from species such as the American white oak, which was visibly affected by its disappearance. Today, management of many American forests involves controlled fires to renovate germination of certain species and to increase the vitality of the forest ecosystem.

The return of the passenger pigeon would help with the natural management of these areas, argue experts who work in de-extinction. But it might not be so easy. For Stanley Temple, professor emeritus from the *Department of Forest and Wildlife Ecology* at the University of Wisconsin-Madison, "the vast forests that passenger pigeons require have partly disappeared or, at best, are very fragmented, and one of the bird's foods – the American chestnut – is functionally extinct. In practical terms, in the near future in which it is necessary to act, the extinction is certainly forever." Temple is correct about the American chestnut *(Castanea dentata)*, which suffered a terrible fungal disease in the first half of the twentieth century that destroyed between three and four billion of these trees in a few decades, especially in the Appalachian Mountains, where it formed around 25% of forest mass. Nevertheless, in recent years various initiatives to make the American chestnut resistant to blight, from crossbreeding with resistant specimens to the creation in 2015 of genetically modified ones that are more resistant, are starting to bear fruit and their replanting has begun in areas where it was prevalent.

The Ecological Role of the Mammoth on the Mammoth Steppe

In the case of the mammoth, restoration of the Mammoth Steppe – substituted by unproductive tundra after its disappearance – would help control the emission of greenhouse gases that are being massively emitted as climate change melts the Siberian land. But, what is the mammoth steppe and how can it be restored?

During the Ice Ages that took place in the last million years, Eurasia and America were covered in ice caps of a thickness comparable to what we today find in Greenland, and which advanced towards the south in the glacial maximums and retreated towards the Arctic and towards the large mountain chains in the warmer periods. Vast extensions of territory, from western Europe to Siberia and Canada, and from the Arctic towards northern China, made up an enormous ecosystem that covered almost half of the earth's surface. This ecosystem, which has been named the Mammoth Steppe (using the word mammoth as an adjective), was made up of a cold and arid steppe dominated by grassy vegetation that nourished large herds of herbivores such as mammoths, bison, woolly rhinoceros, horses, moose, reindeer, muskox and saiga antelopes, and the resulting predators who fed on the herds, such as wolves and cave lions.

After the Last Glacial Maximum, around 18,000 years ago, the climate became warmer and the Mammoth Steppe began to disappear, and with it a large part of its inhabitants. We have already seen how some, such as mammoths, were hunted by humans and how this probably contributed to their disappearance. But the fact is their food also diminished, in parallel with an unprecedented transformation in the landscape that turned the Mammoth Steppe into present-day tundra, taken over by unproductive and swampy land, full of moss and dotted with bushes. Only moose and reindeer knew how to adapt to the change, probably because they could partly feed themselves on moss and various types of bushes; others, like the bison, disappeared from Eurasia but proliferated on the central plains in North America, where they created herds with millions of members that sustained the Native Americans for thousands of years. Others, like the saiga, a strange-looking antelope formerly abundant in all of Europe and North America, did not disappear, but was marginalized to cold areas of central Asia, especially in Mongolia, Kazakhstan and Uzbekistan.

9. The Mammoth Steppe

Some theories maintain that progressive change in the ecosystem eliminated the megafauna from the Mammoth Steppe, but other experts believe that in fact it was the absence of this megafauna, hunted to extinction by humans, that brought about the ecological change. Among the defenders of this last viewpoint is Sergey Zimov, a Russian biologist from the Northeast Science Station in Cherskii, in the Sakha Republic in northeast Siberia. For Zimov, it was the megafauna that maintained the Mammoth Steppe with their continuous grazing, trampling on the bushes and mosses, and fertilizing the land with their excrement.

The grassy ecosystem uses large quantities of water to grow rapidly, while at the same time drying the land, which impedes the water from accumulating in the undersoil as ice. If the herbivores do not take action, the grass dries and accumulates on the surface, but the cold does not allow it to decompose. Consequently, the land becomes unproductive and the nutrients are so scarce that only mosses, which have no roots, can grow on it; the proliferation of moss leads to the retention of humidity and waterlogging under the surface. With the cold, this water freezes and forms permafrost. Something similar could happen in the future with the African savannah, a grassy steppe maintained by the continuous grazing of large herds of herbivores (who at the same time provide sustenance for diverse carnivores). We can imagine the Mammoth Steppe as a type of cold savannah, with its enormous herds of large mammals; a sight that could be shocking for a continent like ours.

It is, in a certain way, an amazing and fundamentally simple idea; the fauna itself takes care of maintaining the ecosystem that sustains it, given that it exercises continuous pressure on the growth of alternative vegetal species and ensures the cycle of nutrients. In their absence, a change is inexorably produced leading to the impoverishment of the ecosystem.

Something similar was detected in New Zealand where, until they were exterminated by human beings around four hundred years ago, some giant flightless birds, the moas, distantly related to ostriches and kiwis (currently the symbol of that country) lived without nearly any competition. Upon their arrival in New Zealand, European botanists observed that numerous species of native bushes had similar forms, such as a very reduced number of leaves or spiny protectors on the edges of the plant, although they had no evolutionary relationship. It looked as if some kind of selective pressure had forced the bushes on the island to adopt these forms. They had found coprolites (that is, fossilized dung) from moas and were able to study their diet. We know that moas put great pressure on the vegetation in New Zealand, which no longer happens, permitting these bushes to grow freely (although they maintain these defensive strategies because they do not "know" that the moas have disappeared).

Zimov's starting point may seem ridiculous, but it is perfectly compatible with scientific method; it has to do with proving that reintroduction of part of the disappeared fauna could turn the current tundra back into the ancient Mammoth Steppe (incidentally, this makes us responsible not only for the disappearance of the mammoths, but also for the disappearance of their ecosystem; perhaps it is starting to be a lot of responsibility on us). In 1989, Zimov chose a territory of around 100 square miles (that he baptized, somewhat provocatively, Pleistocene Park), formed by the bends of the Kolyma River, south of Cherskii; there is also more than 370 square miles of land nearby, which acts as a cushion area for the park and where it could expand in the future. Zimov began by reintroducing a small herd of wild horses and some moose, letting them act freely over the land; currently, he is negotiating reintroduction of some bison, which would be transferred from Canada (given that the Eurasian variety is extinct). As the ecosystem transforms and becomes more grassy, the park will expand its limits and the herds will grow in size. Eventually, Siberian tigers (who are adapted to surviving in winter temperatures as low as 25 or 30 degrees below zero) will be able to be introduced in the park, acting as carnivores equivalent on a trophic level to the disappeared cave lions.

You could object that Zimov is manipulating the landscape, transforming it artificially (although through natural methods), but the truth is that human beings have done precisely this for thousands of years. We are used to seeing phenomena related to climate change in the news, but for now this transformation is nothing compared with what we created with the arrival of the Neolithic and the development of agriculture, around ten or twelve thousand years ago. The landscape was modified progressively, first with deforestation and substitution of ecological diversity with a few vegetal species and selected animals. Domestication implied the progressive selection of traits that were beneficial for farmers (such as cereal varieties of greater size or the fact that mature sprigs do not fall to the ground) and, as such, progressive indirect genetic modification in respect to their wild ancestors.

There is no turning back from many of these changes brought about by humans; the impoverishment of the land in the Middle East from the effects of thousands of years of intensive agriculture has led to the irreversible desertification of these climatically delicate lands. Deforestation of wide areas of the Amazon and central Africa from practicing unsustainable agriculture for only a few years, until the subsequent cleansing of the nutrients from the land due to torrential rains, has a similar effect to what was seen during the Neolithic transition. We cannot argue, then, about the dangers of human intervention in nature, because the current environment is already a product of this intervention in the past.

Pleistocene Park and Climate Change

For Zimov, Pleistocene Park has some additional advantages to help stop climate change. The Mammoth Steppe has a higher albedo (the amount of solar energy reflected off the earth and back into the atmosphere) than tundra, which could consequently lower global warming (objects with lower albedo, such as bodies or dark clothing, tend to heat up more).

Furthermore, the progressive melting of permafrost on the Siberian tundra will allow decomposition of microorganisms from the large quantity of organic material trapped in the Siberian ice, which will release massive amounts of CO_2 into the atmosphere (if the decomposition happens in the presence of oxygen) and methane gas (if it takes place in anaerobic conditions). It has been calculated that the total carbon accumulated in the permafrost could reach 500 gigatons (a gigaton is one billion tons), two and a half times the total accumulated carbon in the biomass from tropical jungles. According to a study by professor Jeff Chanton and collaborators from the University of Alaska-Fairbanks, methane gas emissions into the atmosphere coming from Siberian lakes have increased 58% from 1974 to 2000, which seems to indicate that the process has already begun. These scientists warn that if the temperature continues to increase, up to 2,000 gigatons of methane could escape into the atmosphere, which would escalate the greenhouse effect. In contrast, Zimov argues that progressive development of a grassy vegetal cover in Siberia could partly stop the decomposition of carbon trapped in the permafrost, and in this way palliate the harmful effects described. The presence of large herds of herbivores means they would move aside the snow to get at the grass underneath, as such exposing the ground to the winter cold, which would stop the permafrost from melting and the carbon would decompose.

What would be the role of the revived mammoth in this ecosystem? In truth, the mammoth is not necessary, or it would be necessary for, let's say, purely aesthetic reasons. The Mammoth Steppe could function without the animal that gave it its name with only the other herbivores, such as reindeer, bison or horses, who are still among us. Experts in de-extinction could argue, however, that large herds of mammoths would do the work much more efficiently than the rest of the herbivores together.

Logically, Zimov's viewpoint on ecological change has had its detractors, who discard it as simplistic. For some, the disappearance of the Mammoth Steppe cannot be explained by the disappearance of the megafauna, because the affected territory is enormous and the herds, although large, must have been limited. For the same reason, they criticize Zimov's experiment; it is impossible, they say, to have perceptible effects with so few animals and such a limited area of land, and the current economic situation in Russia does not allow for optimism on the final scope and duration of this experiment. Some ecologists also believe that the grassy ecosystem cannot be reestablished unless it is accompanied by a change in climate, given that current Siberian conditions would be too humid to sustain it. What we are discussing, in the end, is the dynamism of ecological changes and the possibility of being able to intervene in them on human timescales (and not in dozens of thousands of years).

The ability to plan, which exists thanks to the human brain and is, as such, also a product of evolution is, as Dawkins would say, something completely new on this planet. What underlies this debate is the fact that in the future, whether we like it or not, all the ecosystems on earth will be directly managed by human beings. In practice, they already are; often, even on natural reserves, we eliminate invasive species, monitor the reproduction of protected species, plant trees, ensure water resources, etc. It is possible that reinvented species, like natural ones, find themselves in the dilemma of adapting or becoming extinct, and as such we should flee from static conceptions of nature. Nature changed in the past, and will keep changing in the future.

In truth, it is possible that the distinction between natural and artificial species will start to dissolve in the future, when increasingly more transgenic organisms exist that have been designed to survive and be better adapted to changing environmental conditions or new epidemiological challenges. Our role will increasingly be as an administrator of universal diversity and, at any rate, the debate is whether we will be capable of doing this job successfully.

Fear of being Frankenstein

In the famous novel by Mary Shelley, *Frankenstein or The Modern Prometheus* (1818), we are introduced to a scientist, Victor Frankenstein, whose obsession with knowing the secrets of life leads him to create a body made of parts from cadavers, which he brings to life through an experiment with electricity. But upon seeing his monstrous creation alive, he flees horrified from the laboratory. Later, the monster, who has secretly learned to speak and is conscious of his uniqueness, asks the scientist to create a companion for him to alleviate his loneliness. Although at first Frankenstein agrees, he later changes his mind and destroys his new creation before she is finished. Blind with rage at his creator and looking for revenge, the monster first kills the scientist's best friend and later his bride on their wedding night. Frankenstein decides to terminate his creation, and chases the monster to the Arctic Ocean; there the doctor is rescued from among the icebergs by a boat, but later succumbs to exhaustion. The captain, who has listened to this terrible story from the lips of the dying man, is approached by the monster, who tells him that he is going to put an end to his miserable existence. After its publication, Frankenstein became a universal success, as it explores the responsibility of becoming a creator, something that, until science arrived, was reserved for God. The rebellion of the creation towards its creator is a metaphor for the possible disastrous consequences that can come from improper use of technology, and of the punishment that awaits the irresponsible scientist.

There is no doubt that the message from Frankenstein, more up-to-date than ever, hovers in these same terms over the debate on de-extinction. Is it fair that we take over the role reserved for nature, which some people perceive as a knowing force? Is it fair to resurrect species that the natural order "has decided" to eliminate, although it has happened because of instrumental intervention by human beings? Will there not be a big price to pay for this type of technological arrogance?

In the film *Jurassic Park*, a scientist in chaos theory, played by Jeff Goldblum, tries to convince the park owner that bringing dinosaurs back to life is a bad idea: "God creates the dinosaurs. God destroys the dinosaurs. God creates man. Man destroys God. Man creates the dinosaurs." (His companion, played by Laura Dern, jokingly ends the argument: "The dinosaurs eat man. Woman inherits the earth.")

The Moral Debt to Extinct Species

Conversely, a possible counter-argument would be that we have the moral obligation to bring extinct species back to life. From a philosophical viewpoint, this only makes sense if we consider that humans have a moral obligation towards "something" that has its own rights, which we humans have violated. Which leads to the crucial question of if animals have rights (that is, if they have a legal or moral legitimacy to have or obtain something). If animals do not have rights, then we humans do not owe them anything. There is no agreed upon response to this question among experts in bioethics; some think that if animals feel pain, then they have rights, and others believe that they would only have rights if they show that they can make moral decisions. Independently of this debate, it could be that humans feel sufficiently guilty to try to make amends, at least in part, for the damage inflicted in the past on species that have disappeared because of us.

For Paul R. Ehrlich, the debate is tainted by what he considers a "species-centric" view of diversity, reinforced by hundreds of sterile discussions on how to define a species, which, as we have said in earlier chapters, is clearly an arbitrary concept. For Ehrlich, it is as if geologists worried about how to define what is a mountain and went no further than that. Conservationists, he argues, are also trapped in this debate and do not realize that the critical problem is not the extinction of species, but rather the disappearance of populations, which are the natural entities that maintain ecosystems and allow the creation of new diversity (in an extreme case, it is clear that the disappearance of all of a species' populations will lead to its extinction).

Be that as it may, the fear of being Frankenstein is not only about irrational arguments related to the fear of modifying the established order, but is also about purely scientific objections formulated regarding the risk that our intervention ends up creating an ecological disaster.

The Fear of Ecological Disaster

Experts in biological invasions know that the effects of introducing new species in other ecosystems are often catastrophic and unpredictable for the native species. One of the most well-known examples is the introduction of rabbits in Australia with the arrival of the first colonists in 1788. The plague, however, appears to have originated with the liberation of 24 specimens by Thomas Austin, a colonist who liked to shoot them for fun, in 1859. The rabbits ended up being so prolific that it is calculated that by 1950 there were around 600 million. For decades, the Australian authorities sought different strategies to try to control the plague that devastated the pastures and crops, from the construction of an enormous fence along the continent to try to contain them to the intentional introduction of a virus, the source of myxomatosis, to decimate them. Even so, it is calculated that the population continues being between 200 and 300 million individuals. It is not a unique phenomenon; in Hawaii they counted 22,070 species of all types of organisms (9,805 of which are native to the islands), and of these 4,373 are invasive species.

The individuals brought back to life would be like invasive species in their own areas of original distribution, because these ecosystems would not be exactly the same to the ancestral ones, and because the trophic chains would have readjusted and adapted to the absence of the extinct species. The return of the species that had disappeared could leave it in competition with others that were occupying their ecological function, and were just as native to the place as the de-extinct species. There is no certainty that the de-extinct species would return and simply take on its previous ecological role as if nothing had happened. Thinking that if one of these species becomes a problem we would simply exterminate it again, just like we did the first time, seems monstrous from a moral point of view.

In any case, it is evident that any reintroduction would require a contingency plan, and it would be necessary to also determine who would be economically responsible for possibly eliminating the new species should it become a problem. For some experts in biological invasions, we would need to consider not only why we should bring the passenger pigeon back to life, but also why we should bring it back now and not within 50 years; when all is said and done, they argue, it can wait a few more years because, in fact, it is already extinct. Perhaps prudence recommends better planning for possible contingencies – if it is possible to plan for the unknown – instead of jumping in to do it immediately. I believe this to be a poor argument when someone has the ability to do something right now. In truth, legislation does not exist on this subject at the moment. Scientific advances move so rapidly that synthetic biologists could liberate a de-extinct species without having to follow any regulations. The only legislation that could adapt quickly to the problem would be that which regulates transgenics.

The Question of Resources

Another fear about de-extinction is that resources would be allocated for these initiatives that otherwise would go to preventing future extinctions. On occasion I have stated to certain media that it would be "inadmissible to allocate resources to revive extinct species when these resources could be used to save some species that still exist and are in danger of extinction." But I have come to understand that, although it has a moral basis, it is a fallacious argument. Research funds are not being taken from the salaries of guards who fight against poachers in African parks, nor are they contributing to promote illegal trafficking in ivory. They are different things and are not connected. Nor should the fact that de-extinction initiatives are very expensive be frivolously criticized; these types of projects have the advantage of putting diverse leading scientific institutions run by highly-qualified people in contact, and are interesting to investors who otherwise would not ever think about getting involved in environmental conservation projects. The science created in cutting-edge work always has applications in other fields, and is one of the engines of progress.

Obviously, this does not mean that all efforts and resources possible should not be put into conserving endangered species, but the responsibility of stopping processes cannot be conferred on scientists who explore new fields. In my view, Ehrlich himself offers a mistaken argument by affirming that activists in favor of women's rights are doing much more for global diversity than biologists who work on de-extinction (since in countries where these rights are established, they lead to lower birth rates); it is erroneous and unjust. Research and knowledge, like art, are justified in themselves, and scientists in de-extinction cannot be morally disparaged for not dedicating themselves to other more applied work that, additionally, is not part of their competency.

It is precisely the results of de-extinction studies that can contribute indirectly to conservation. And perhaps in a more practical, as well as imaginative, way. For example, with de-extinction techniques it would be possible to restore lost genetic diversity in a species in danger of extinction that has such a low demographic efficiency that inbreeding and the loss of resulting diversity compromise its future viability. Paleogenomics could trace which genetic variation has been lost in the last thousands of years in this species, and synthetic biologists could reintroduce it in the limited survivors.

The Case of the Iberian Lynx

The complete genome sequencing of the Iberian lynx (*Lynx pardinus*) in 2016 revealed an alarming loss of genetic diversity, as well as the accumulation of harmful genetic variants that brought about lowered viability in the scarce current population, with lower-quality semen, testicular disorders and juvenile epilepsy. Although one of the possible actions to increase genetic diversity would be to breed current specimens with those of the Eurasian lynx (*Lynx lynx*), an alternate possibility would be to edit the genome and manually eliminate the genetic mutations most harmful for their survival. DNA sequences have been retrieved from specimens dating to 50,000 years ago from the entire Iberian peninsula, so it would be possible to also explore what diversity has been lost in the three large bottlenecks detected in their recent evolutionary history (one nearly 47,000 years ago, another 300 years ago and the other in the twentieth century).

There could also be the case of a species that becomes threatened, for example, by a new type of pathogen. Recently we have seen how the Tasmanian devil seemed condemned to disappear because of a rare type of contagious cancer on its face. Synthetic biologists could modify one or various genes to make this species resistant to the new adaptive challenge. In a few generations, the resistant individuals would have substituted those susceptible to contracting the disease, and the danger would be overcome thanks to human actions.

Likewise, genetic engineering could be done on species destined to disappear from their natural habitats due to the effects of climate change or the impossibility of ensuring effective conservation policies in their countries of origin. The modification of some key genes could allow the last Sumatran rhinos to survive at higher latitudes and in colder climates, and as such they could create reserves farther north, in countries with greater economic or social stability. It would not even have to be the same habitat, given that aspects of their physiology and metabolism could be modified so that they adapt to ingesting new types of plants. Perhaps they would no longer be one hundred percent the original rhinos, but at the same time we would have been able to save them from extinction with minimal intervention from genetic engineering.

A Message of Hope

To a large extent, conservation is based on hope; we invest money in trying to preserve threatened species with the hope that they do not become extinct and, even, that they can recover. I believe it would be a mistake for conservationists to perceive the birth of de-extinction as a threat, because in part it offers a glimmer of hope that should not be underestimated. The resurrection of charismatic species could be used as a launching pad to promote comprehensive conservation in new territories. At the same time, it is true that de-extinction could be misunderstood at the grass-roots level, giving the erroneous impression that an extinction is not as worrisome a phenomenon – and in its own way, terrifying – as in truth it really is. Probably, a suitable viewpoint would be to contemplate de-extinction not as a frivolous competitor that takes control of resources, but rather as one more step in conservation policies. Some argue, for example, that the case of the northern white rhinos we mentioned in previous chapters is already more an initial case of de-extinction than conservation.

With everything we have seen throughout this book, it is evident that extinct species, as they were in their original form, will never return to life. We will never again see in action these specific combinations of genes. They will be genetic chimeras that have a component, which may or may not be a majority, from the original species. But above all, their presence in the twenty-first century will be different because the world is also different. Which is why we should not think about de-extinction as restoration of something that is ancient, but rather as the beginning of something new.

Conclusions

For centuries, philosophy has debated the role of human beings in the natural world, at times putting them above the rest of the animals and at other times equating them to our closest living relatives, the chimpanzees. But this is a debate that, to a certain point, is obsolete: whether we are one more part or an exceptional piece of nature, we find ourselves in a critical situation of having to make decisions about it – and about us. Human beings now hold the power to modify the global ecosystem and have the duty to manage it correctly.

This management job swings between finding the necessary resources to feed and assure the well-being of a growing world population and at the same time preserving biodiversity. Although both needs may often be in conflict, it is increasingly clear that they are strongly interrelated and that we should consider them jointly and rationally; the global ecosystem constitutes life support for the future human population. Public opinion towards this problem is considerably polarized and ranges from radical conservationists to extreme "exploiters" of resources. This disparity is curious, and is not seen in other problems having a biological basis, such as animal rights, racism or discrimination against women. One of the pending tasks for globalization is to create an ethical awareness of the earth that searches for how to blend environmental conservation with economic development.

Science is no longer just a useful tool to understand the natural world, but rather is going to be necessary to transform it. In such a changeable world like the current one, the first people who should adapt to change are naturalist scientists, because we are not so much in times of diagnosis but rather in times to act. And that is why we are going to need many of the techniques that de-extinction is based on: cloning, genome editing or synthetic biology will be widely used in a few years. It is obvious that mistakes will be made in the process of transforming the global ecosystem, but we will not be in a worse situation than the one inaction would produce.

I have the perception that, in a few decades, the current distinction between non-genetically modified organisms and genetically modified organisms will no longer be applied. Many humans and many domesticated animals and plants will have been edited, the first for medical ends and in the latter, to improve production. In a few more years, synthetic organisms will be designed so they can take charge of "terraformation," that is, the transformation of the terrestrial ecosystem designed by humans to combat challenges, such as climate change, on a planetary scale. Synthetic organisms would be able to help capture excess carbon dioxide, eliminate plastics and petroleum derivatives or to increase the resilience of ecosystems to catastrophic changes. As the expert in complex systems Ricard Solé says, "it could be that to save the planet, we will have to redesign it."

In this future framework, de-extinction, as we now perceive it, will not be frivolous or a science-fiction fantasy, but rather one more tool. I do not know if we will see the passenger pigeon or Xerces blue butterfly fly again, but it is probable that organisms similar to some extinct species will have been "designed" with the goal of restoring certain ecosystems or making them more dynamic or resilient. It is also probable that species in danger of extinction will have been modified to facilitate their future viability, making them more fertile and more adaptable to changing climate conditions. The division into categories itself that current terminology implies perhaps will not exist, because it will no longer make sense. And it is possible, then, that we will no longer speak about "de-extinction," but rather of "bio-design."

Further Reading

Barrow Jr, Mark V. (2009). *Nature's Ghosts: Confronting Extinction from the Age of Jefferson to the Age of Ecology*. University of Chicago Press: Chicago and London.

Church, George M. (Ed Regis) (2012). *Regenesis: How Synthetic Biology Will Reinvent Nature and Ourselves*. Basic Books: Philadelphia.

Darwin, Charles (1868). *The Variation of Animals and Plants Under Domestication*. 1 ed. John Murray: London.

Diamond, Jared (1997). *Guns, Germs, and Steel: The Fates of Human Societies*. W.W. Norton & Co: New York.

Dugatkin, Lee Alan; Trut, Lyudmila (2017). *How to Tame a Fox (and Build a Dog)*. University of Chicago Press: Chicago.

Lister Adrian; Bahn, Paul G. (2007). *Mammoths: Giants of the Ice Age*. University of California Press: Berkeley, California.

Paddle, Robert (2002). *The Last Tasmanian Tiger: The History and Extinction of the Thylacine*. Cambridge University Press: Cambridge.

Shapiro, Beth (2015). *How to Clone a Mammoth: The Science of De-Extinction*. Princeton University Press: New Jersey.

Wilson, Edward O.; Peter, Frances M. (Eds) (1988). *Biodiversity*. National Academy Press: Washington DC.

Wilson, Edward O. (2002). *El futuro de la vida*. Galaxia Gutenberg; Círculo de Lectores: Barcelona.

Zimmer, Carl; Emlen, Douglas J. (2013). *Evolution: Making Sense of Life*. Roberts and Company Publishers Inc: Greenwood Village, USA.

Available soon:

Populisms
Carlos de la Torre

Happiness
Ben Radcliff & Amitava Krishna Dutt

Human Rights
Peter Rosenblum

French Revolution
Jay Smith

Global Waste
Anne Berg

Federalism
John Kincaid

www.ingramcontent.com/pod-product-compliance
Lightning Source LLC
Chambersburg PA
CBHW060031210326
41520CB00009B/1090